COUNTY FAIRS

WHERE AMERICA MEETS

COUNTY FAIRS

WHERE AMERICA MEETS

TEXT BY JOHN McCARRY
PHOTOGRAPHS BY RANDY OLSON

The world's largest nonprofit scientific and educational organization, the National Geographic Society was founded in 1888 "for the increase and diffusion of geographic knowledge." Since then it has supported scientific exploration and spread information to its more than nine million members worldwide.

The National Geographic Society educates and inspires millions every day through magazines, books, television programs, videos, maps and atlases, research grants, the National Geography Bee, teacher workshops, and innovative classroom materials.

The Society is supported through membership dues and income from the sale of its educational products. Members receive NATIONAL GEOGRAPHIC magazine—the Society's official journal—discounts on Society products, and other benefits. For more information about the National Geographic Society and its educational programs and publications, please call 1-800-NGS-LINE (647-5463), or write to the following address:

National Geographic Society
1145 17th Street N.W.
Washington, D.C. 20036-4688
U.S.A.

Visit the Society's Web site at
http://www.nationalgeographic.com

Copyright ©1997 National Geographic Society
Photographs Copyright ©1997 Randy Olson
All rights reserved. Reproduction of the whole or any part of the contents without written permission is prohibited.

McCarry, John.
    County fairs : where America meets / text by John McCarry ; photographs by Randy Olson.
        p.   cm.
    ISBN 0-7922-7091-6
    1. Agricultural exhibitions—United States. 2. Fairs—United States. 3. Agricultural exhibitions—United States—Pictorial works. 4. Fairs—United States—Pictorial works.   I. Olson, Randy. II. Title.
S541.M35 1997
630'.74'73—dc21   97-11868
                                                                    CIP

Half Title Page *Fair time comes to the city, too—in a big way. The 487-acre Los Angeles County Fair, Hotel and Exhibition Complex (Fairplex) boasts more than 350,000 square feet of indoor exhibit space, a racetrack with a 10,000-seat grandstand, and 12 acres of carnival grounds.*

Title Page *Families enjoy treats, neighbors salute each others' skills, and communities honor local heritage at county fairs across the United States. Visitors to the Clay County Fair mingle among scores of food stands in a straw-slurping, snack-sharing celebration of Iowa's abundance.*

Dedication

For my parents.

John McCarry

For my father, whose farmboy resourcefulness, photographer's eye, and writer's insight helped shape my view of the world.

Randy Olson

1

2

3

4

5

6

Contents

CHAPTER 1

Fair Time

On a bright August afternoon, Flo Bowen, retired Navy nurse, stood on the neatly trimmed lawns of the Kitsap County Fair in Bremerton, Washington, raffling off a glinting maroon Town Car. Before her— lolling on the grass or strolling on the blistered asphalt or passing in and out of the exhibition halls, barns, and tents—were the people of Kitsap County. It was a Pacific Northwest crowd, where cowpoke meets grunge, where Stetsons and pointy-toed boots are as common as tattoos and nose rings. It was a county fair crowd, in which the old, the middle-aged, and the young effortlessly came together and were happy.

"Well," Flo Bowen said, selling her tenth raffle ticket since morning. "Standing here you definitely see that it takes all kinds. You see 'em all."

Fair time: That moment between summer and fall when the dizzying crush of an urban crowd creates itself in the emptiness of the countryside. It is harvest time, when proud farmers from across the county bring in their fattest pumpkin or their prettiest ears of corn or their juciest tomatoes and are judged against the efforts of their neighbors. It is that time

It's hard to be patient when you've waited all year for the fair and you're ready to shine. At the Brazoria County Fair in Angleton, Texas, young girls decked out in stars and sunflowers are poised to parade their talents before families, friends, and neighbors, who are just as eager to applaud.

when kids across the country lose themselves in the fantasy world of the carnival.

Today, less than 3 percent of Americans are directly engaged in farming. Yet Americans continue to love attending county fairs. One reason, of course, is that fairs offer an occasion to celebrate our agrarian traditions and to showcase our values as a nation—values like family, hard work, and inventiveness. More importantly, they allow us, as communities, to come together, face to face, and get to know one another.

A county fair is many things to the many people who attend them, not least of all a chance to glimpse the American past. Yet county fairs have not endured by being annual historical reenactments, but by evolving as American society evolves. As the nation becomes increasingly urbanized (the 1990 U.S. Census informs us that slightly more than 75 percent of Americans live in urban areas), and, as a result, more standardized and monolithic, the need to feel part of a regional community is more urgent now than ever.

In the summer and fall of 1996 I crisscrossed the country, visiting county fairs from Alabama, California, and Iowa to Massachusetts, Texas, and Washington. Whether in farm country or in areas with more diverse economies, almost all of these fairs thrived as local people got together to reward one another for their small accomplishments, to celebrate their subtle regional differences, and to gently compete. If a fair had momentarily faltered, it was perhaps because it was struggling to understand what sort of community it had become.

Today, the fairgoers are just as likely to see 4-H lambs and patch quilts as they are to see demolition derbies and fast-talking salesmen trying to sell a set of Ginsu knives or a computer. Whether they are there to compete and educate or simply to buy and sell, Americans continue to come to county fairs in an effort to be together.

Right A century ago a parading brass band and 83 yoke of oxen opened the Tunbridge Fair. These days stock showing, pony pulling, midway glitter, and big-name grandstand acts fill this quiet Vermont valley with bumper-to-bumper crowds.
Following pages *In Amidon, North Dakota, the Slope County Fair Rodeo offers $1,000 in purses for the cowboys and cowgirls who triumph in its saddle bronc, steer wrestling, goat tying, bull riding, calf roping, and other competitions. Blending community traditions with entertainment spectacles, fairs large and small thrive across the country.*

Almost all world cultures have organized fairs, from the Chinese in the 12th century B.C. to the Incas in the 15th century A.D. The earliest fairs in the Western tradition were likely those of the Phoenicians, which were organized exclusively around commerce. In ancient Greece, fairs evolved from purely commercial gatherings into events honoring various gods. While under the Roman empire, fairs—the English word comes from the Latin *feriae* or "holy days"—remained religious and commercial events. It took a long time for long-distance travel to be restored after the fall of the Roman empire, but by the 13th- and 14th-centuries fairs were again organizing throughout Europe, as well as in the Middle East, most notably in Mecca, where they were earlier held. Some medieval fairs were established for the sale of one particular commodity, such as livestock or cattle or cloth. But almost always, entertainment like fire-eating and tightrope-walking was also provided by individuals or troupes. By the 18th century, fairs in Europe began to lose much of their commercial importance as the culture of shopkeeping evolved and transportation became easier.

The history of European fairs in the New World began in the 1620s in New Amsterdam, when the Dutch settlers decreed

that there should be two fairs a year held in the colony: one in October for the sale of cattle, the other in November for the sale of hogs. These market fairs were patterned after the *kermis* fairs of Holland, which were boisterous events that included clowns and acrobats and games like "pulling the goose," in which young men competed to see who could catch a goose covered in grease.

While fairs, such as those held in New Amsterdam, were essentially European fairs reenacted upon New World soil, the first authentically American fair was the agricultural or county fair. At the time of the country's independence, the United Stares was overwhelmingly agrarian. Most Americans continued to cultivate the soil in the old colonial way, having adopted native crops like corn and tobacco as well as continuing to raise crops and livestock their forefathers brought from Europe. With independence came the need to innovate agricultural production while preserving soil fertility, and thus secure economic independence from Europe.

In order to introduce European farming innovations to the United States, patriotic gentlemen in America's port cities formed agricultural societies in Charleston, South Carolina, and Philadelphia in 1785. Two more societies were founded in New York and Boston in 1791 and 1792. These gentlemen farmers, if farmers at all, advanced possible schemes that might help the United States achieve economic self-sufficiency, such as the domestication of the moose.

Just as zealous, though somewhat more practical, in this ambition was Elkanah Watson, a gentleman farmer and one-time revolutionary. Born in Plymouth, Massachusetts in 1758 of Mayflower descendants, Watson was apprenticed to merchant John Brown of Providence, Rhode Island early in the War for Independence and carried government dispatches to Benjamin Franklin and John Adams in Europe in 1779. As he established

his own trading business in France, Watson traveled across Europe over the next five years and recorded his observations on European manners, morals, farming, industry, and canals.

After retiring from business in Albany, New York in 1807, Watson returned to his native Massachusetts, where he bought an estate outside of Pittsfield, a town in the extreme western part of the state. Here, Watson exhibited on the village green in 1808 two Merino sheep, a Spanish breed valued for its fine fleece, that he had purchased from Chancellor Robert R. Livingston of New York, the visionary who had earlier suggested domesticating the moose. Watson hoped not only to interest local hillside farmers in the breed in order to guarantee a steady supply of raw wool for his newly established woolens factory, but also to support American manufactures. The surprising success of that 1808 exhibition gave him an idea.

Two years later, Watson convinced local farmers to hold a larger livestock exhibition. Its success led to the establishment of the Berkshire Agricultural Society the following year. Elkanah Watson was its president. The society was organized for the sole purpose of holding an annual county fair, and its first one on September 24, 1811, was a festive occasion. Watson organized an impressive parade that included a band of music, floats carrying a broadcloth loom and a spinning jenny,

and 60 yoke of oxen drawing a plow held by two of the oldest men in the county.

A decade later, Watson would write of the county fair, and of his part in its conception, "The grand secret in all our operations was to trace the windings of the human heart." In 1811 Watson traced one of the human heart's tautest windings. He offered prizes worth $70 for the best livestock in the county. More than 3,000 people attended this first Berkshire County Fair.

Two years later in 1813, Watson expanded even further the scope of the county fair, inviting women to contend in the domestic skills of cloth production. He included these competitions in an effort to encourage local households to lessen their dependency on European products. Another original feature of this fair appealed especially to women, although not necessarily to the religious leaders of the community, who had earlier cooperated with Watson. That year the fair closed with an "agricultural ball," at which Watson later reported, "Females are seen ornamented in the work of their own hands and native products of our lands."

Despite, or perhaps because of, such seeming levities, Watson's county fair did indeed serve the sober purpose that he had intended. As other communities began to organize county fairs throughout the young republic, farmers came to attend them in increasing numbers not only to compete with one another for money and also premiums (prizes), but to learn. By the 1840s, county fairs would come to be showcases for new American inventions, such as Cyrus McCormick's reaper and John Deere's steel plow, as well as for imported livestock.

But the county fair also had always fulfilled another purpose. Ever since Watson's day, it had become the social event of the rural year. It had provided a morally legitimate and socially sanctioned reason for farm families to rest from their labors and travel to town to mingle and enjoy each others company.

Top left Pete and Jim Bansen corral cattle they plan to take to the Humboldt County Fair, including one beast that is resolutely uninterested in leaving her familiar Ferndale, California home. Human intelligence meets stubborn mass: Adult cattle routinely weigh more than 1,000 pounds. Bottom left The Coppinis of Humboldt County, California, have raised a dynasty of dairy champions. They preserve the mounted head of one of their first cows. Her offspring produced milk exceptionally high in butterfat. Cream from such rich milk becomes profitable butter, ice cream, and cheese.

As fairs grew in popularity, their scope continued to expand far beyond agricultural edification. By the 1850s, baby shows, in which proud mothers vied with one another for cutest offspring, were organized alongside livestock competitions. Horse racing also became enormously popular and racetracks came to dominate many fairgrounds. While initially a male-dominated activity, women equestrians eventually began to compete in their own events as well. This horrified many, not least of all a contributor to *Country Gentleman* who penned these words in 1857: "Some of our country societies are little better than race courses and trotting matches....Even the modest, well-behaved daughters of our farmers have been induced to enter the listing as riders....God forbid that American females, the daughters of our farmers, should thus pander to the depraved taste and a vicious appetite. The speedy downfall of such societies is certain."

As crowds attended fairs in increasing numbers, many of them just for the thrill of a day at the races, jugglers and acrobats came to divert them and vendors with trays of 19th-century fast food carried atop their heads wove through the crowd. Alarmed by just how convivial some fairs had become, New York State legislators passed an ordinance in 1862 that permitted fair organizers to "...prevent all kinds of theatrical, circus,

or mounbank [sic] exhibitions and shows, as well as huckstering or trafficking in fruits, goods, wares, and merchandise of whatever description for gain on fair days and within a distance of 200 yards of the fairgrounds." The ordinance was largely ignored. The county fair had taken on a life of its own.

I spent the latter part of my childhood in Northampton, Massachusetts, a city of about 30,000 people 60 miles east of Pittsfield. My three brothers and I would eagerly await fair time—my eldest brother, Bill, so he could show his Morgan horses, the rest of us just to spend the five bucks our parents would give each of us for the event. For me, as for most kids, the biggest thrill of all was the carnival—the cotton candy and candied apples and the crazy rides that would spin you around and turn you upside down to the sound of Top 40 tunes. And, of course, the elusive promise of winning the big stuffed animal from the carnie who kept urging, "Just give her one more try, kid, you'll be sure to get it next turn. Thataboy." I never won the big stuffed toy, although each year I would blow my money—25 cents at a time—trying.

One year when my brothers and I had prematurely exhausted our allowances, Bill had the inspired idea of going to our grandmother's house to ask her for more fair money. Grammy was the kindest grandmother any boy could ask for. But, unfortunately for us, she was a dyed-in-the-wool, can't-pull-one-over-on-me Yankee. Her family had come from Holland to New York when it still belonged to the Dutch but eventually moved to Pittsfield, where she was born in 1890. Grammy knew all about spendthrift boys and county fairs. "Didn't your father already give each of you five dollars?" she asked, her pale blue eyes unwavering. We all looked at our shoes. We didn't get the money—but we each got a kiss and a slice of homemade apple pie.

For the first time since I was a kid, I returned to the Three County Fair in Northampton, Massachusetts. This fair was

originally held in 1818, less than a decade after Elkanah Watson's first fair. I bought a ticket—entry alone was now five dollars—from Gordon Shoro, a local man who has worked at the fairgrounds the past couple of years. I asked him why he decided to help out at the fair. "It takes me back," Shoro replied in a quiet voice. "I see the kids and their faces and how excited they are about coming to the fair, and I remember coming here when I was 10 years old, with only 52 cents in my pocket. I used to sneak in through a hole in the fence and then run like heck."

In many ways county fairs have changed little from the time when youngsters snuck in under the fence. At the Cullman County Fair in Cullman, Alabama, the biggest poultry producing county in the United States, agriculture continues to take precedence over all other things. Chester Freeman, a gentlemanly Southerner and one of the 11 members of the fair board, told me as we strolled through the modest-sized fairgrounds, "The whole idea is to take you back to the heart of what is Alabama."

While the Cullman County Fair only began in 1954, it is in most regards very similar to the fairs of a century ago. As Freeman said, "Competition among neighbors is an important part of what goes on here—you know, the whole idea of this year is I'm gonna make a better apple pie than that ol' gal who won last year. We have people here who've been entering exhibits—and winning ribbons—for each of the 42 years that we've had the fair."

Gathered with five generations of her family, flag-bearer Mureen Walker says Oregon's Curry County Fair began as "a fat lamb show to teach ranchers the finer points of getting the highest quality production." Mureen's mother Rose and 98-year-old grandmother Ruby attended that first fair as well.

Half a century ago folks along southern Oregon's rocky, isolated coast traveled largely by boat, horse, and foot. Today, Highway 101 will take you through Pistol River north to Gold Beach, home of the Curry County fairgrounds, but a horseback ride along the beach is still a beautiful option.

Freeman told me that that year they were awarding a total of $8,000 in prize money—a far cry from the $70 worth of premiums Elkanah Watson awarded at the Berkshire County Fair back in 1811. Nonetheless, Freeman maintained, "People still don't value prize money as much as they do that ribbon."

We paused at a concession stand on the fairgrounds, which, like the fair itself, was run by the local Lions Club. Freeman bought me a cup of coffee and I asked him about the daunting task of organizing a fair. When does the fair board begin? "Well, in November we elect the new officers; then we meet with the cattleman's association so they can critique the past fair for us; then we get together with the extension agents...." His voice trailed off. He turned to me and a smile formed on his lips. "Well, when it all really starts, the true beginning of any fair is when those first seeds get put in the ground. It's when the farmers plant their best pumpkin or corn seeds in that special corner of their choicest field so that it'll all come out just right for gathering at fair time."

While agriculture still finds its showcase on all county fairgrounds, fairs have become places not so much to learn how to improve farming, but to celebrate our traditions and recall our American history. America has changed greatly since the 19th century, and so, too, have our traditions.

One example of how Americana constantly recasts itself is chuck-wagon racing. A century ago chuck wagons would race across the prairies in order to set up camp and make meals for cowboys driving herds of cattle westward. Today, chuck-wagon races provide entertainment for county fair audiences across the American Midwest and western Canada. At the Clay County Fair in Spencer, Iowa, the fairgoers gathering to watch that day's races were dressed like the farmers they were, with dusky blue jeans and mud-caked boots. As the big outdoor arena filled up beneath a vast Midwestern sky, patriotic country rock songs blared from the speakers. The announcer told jokes passed up from the audience. The jests, like the competitions themselves, were harmless ones, in which no one's feelings ever got hurt. My favorite: "What's the best thing to take back from Minnesota? Interstate 35!"

And then the show began, as four chuck wagons, each painted in a different color and drawn by a team of four horses, competed at heart-stopping speeds. First there were the chuck-wagon races themselves, then the chariot races with carioles made of oil barrels on wheels. This was followed by a farmers' race, in which lanky jockeys mounted retired race horses, and, finally, a "powder puff race," in which women took the reins.

While the races provided an afternoon's amusement for fairgoers, it provided men like Don Essick a chance not only to honor the American past, but also to keep alive a family tradition. I met up with Essick at the horse barns after the races, where he and the men who had just raced stood around, shirtless, drinking a favorite beverage jammed into thick rubber sleeves. Resting upon the grass were the deeply colored wagons, all of them farm wagons identical to those used to pick corn in the Midwest of the 1800s. The horses that had pulled the wagons stood side by side in their stalls.

Essick leaned against his wagon, which was painted jet black. A fit man with a farmer's weathered face, he told me, "I first started outriding for my father when I was 12. Eventually, when he thought I was ready, he retired and let me drive. Now my two boys outride for me. I've been driving for 23 years now."

Isn't chuck-wagon racing a somewhat dangerous pastime? "Driving down the road's dangerous, too," Essick said, punctuating the remark with a distinctive chortle. "Mind you, I've taken a few spills in my time. Thing is these farm wagons weren't made for this kind of speed. You gotta be going 35 miles

an hour in a good race, so if you tip over, things get messed up. The only part that's original on this wagon is the back axle. It's a good thing we got Amish in the state of Iowa. They're the only ones who know how to put these things back together."

Essick, who works full time for a seed company, only races his chuck wagon for nine to fourteen days in the summer. Most of the time, he said, these races take place at county fairs. "You can't get the crowd at state fairs," he explained. "I guess in the cities they don't go for chuck-wagon races like they do here."

I asked Essick what it was about chuck-wagon racing that made him want to do it for 23 years. "Well, I don't do much golfing or fishing. I gotta have some hobby," he replied with his characteristic chuckle. "And it's exciting. There's never any race the same. Not everybody can take four horses out there and drive 'em."

But there was also something else that kept Essick coming back year after year to county fairs like the one in Spencer, Iowa. "I want to be the first to drive horses for 25 years," he confessed. "No one else has done it that long, so I'd like to be the one to do that. I only got two more years before I can say I've accomplished that. Then I'll let my boys drive."

In a country where "the bigger the better" has become the yardstick of value, county fairs remain a place to honor smaller human triumphs. Whether they are there to show off a favorite layer cake or a patch quilt or a prize heifer, or, like Essick, to quietly celebrate a more private kind of accomplishment, Americans continue to come to county fairs to show off their best.

The windings of the human heart, I discovered, remain much the same as they did in Elkanah Watson's time. And, as Watson surmised back in 1811, good-natured competition can serve a purpose far larger than itself. Watson wanted to educate the farmers of early America, but he also wanted to unite them.

When neighbors come together to compete at a county fair, what remains important is not so much the contest's outcome, but that they are, quite simply, together.

I recall standing one brightly lit fall afternoon with Arnie Barber, a friendly, white-haired gentleman who was donating his time to grill up sausages and sauerkraut at the senior Kiwanis Food Stand in Spencer. Together, we overheard a middle-aged woman say to her blue-haired companion, "Why, that's Dolly Parkerson over there by the pickup. I'd know her walk anywhere. Let's go over and say hello…." As the women drifted toward the pick-up, and their conversation along with them, Barber smiled and shook his head. "You know, it's funny," he said. "People who might never say hello to one another in their own towns fall all over one another like old friends when they meet at the fair."

This is not only the case in small middle-American towns. Harris County, which is the most populous county in Texas and the third most populous in the United States, revived its fair in 1978 after a century of dormancy. The original fair began in Houston in 1870, but following a yellow fever epidemic in 1878, it moved to Dallas, where it reincarnated itself as the "The Grand Fair of Texas." The Texas State Fair has since become one of the largest fairs in the United States, with a yearly attendance of more than two million. Why, then, the need to resurrect the original fair in Houston?

John Jellison, also known as "Boots," indirectly provided the answer. I met Jellison, who is a member of the Houston

Teenagers cuddle at Alabama's Cullman County Fair. Ribbons and rides aren't the fair's only attractions. "The midway and the dances are wonderful places to enjoy the company of special friends," says Texas fair manager Anita Rogers. "Many young hearts have been stolen at a county fair."

Working eight-hour shifts, the carnies operating this ride at Oregon's Curry County Fair sit in the middle of the whirling mechanism and watch as riders are plastered to the walls by the force of the spin. Some cling desperately while others take advantage of their temporary suspension to try a few tricks.

Farm and Ranch Club that organizes the fair, as he helped out at a Coca-Cola stand. Jellison, who described himself as "a California boy who went country big time," showed me his large, Western-style belt buckle, upon which was embossed the word "Boots." He went on to tell me the story of how he had earned the name.

"When I first joined up with the Houston Farm and Ranch Club," Jellison recounted in an accent that, like the man himself, had long ago made the transition from Californian to Texan, "I was wearin' the boots I'd brought with me from California—candy-ass boots with zippers on the sides. Then one night I got picked on by a pair of Texas twins, who told me if I wanted to be a real cowboy I had to get myself a pair of cowboy boots. That was the night I became a Texan."

It was not just a new name that Boots had earned as a member of the Houston Farm and Ranch Club. He showed me some medallionlike pins affixed to his cowboy vest. There was one identifying him as a club member, another indicating that he was a fair member, and yet another proving that he had completed an arduous trail ride. "You can't buy these," he told me, proudly. "You got to earn 'em."

As we spoke, fairgoers queued up to pay for cups of Coca-Cola with plastic tokens. Many were dressed in cowboy hats and long black dusters that reached down to the heels of their cowboy boots. The only thing that seemed to identify them as the urban cowboys that they really were were the cellular phones dangling like six-guns from their thickly studded belts.

Boots explained to me that benefits from the fair go mostly to sponsoring activities, such as 4-H and Future Farmers of America, and scholarships for Houston's youth. "We're not just a bunch a rednecks out here to have a party," he said as he cheerily handed out Coca-Colas while drinking a cup of beer himself. "This is a party, sure. But it's a *community* party. We

have a goal—to help out the kids. If it wasn't for that, none of us would be here."

He turned to me to catch my eye. "We're like a family—and that's a feeling you can't just get anywhere."

For a few days each year the empty lot or pasture that is the fairgrounds transforms itself into a village. The exhibition hall becomes crowded with big sunflowers and quilts and jars of jam, all painstakingly produced by locals. The animal barns become a menagerie of livestock, all lovingly raised and groomed. The entertainment tent becomes a parade of fair queens and magicians and five-year-old Elvises, the arena a roar of rodeos and rock concerts and stock car races. The midway becomes a honky-tonk fairyland of rides and games and artery-clogging food. And the commercial building becomes a cacophony of pitchmen trying to make a buck and live the American dream.

Along the narrow passageways of dirt, asphalt, or lawn that connect and group together this mythical village roam people of all ages and of all backgrounds. Many have nothing in common other than that they all live within the boundaries of the settlement known as a county. Many come with the express intention of reuniting with "a family"—whether it be the family of sheep raisers, or the family of chuck-wagon racers, or the family of the carnival.

I was struck again and again by the small townishness of a fairground, even when it was located in a big city like Houston. Usually by the end of fair week I would have people approaching me, rather than the other way around, with comments such as, "Oh, yeah, I've seen you around—you're the guy with the notebook." Concessionaires memorized my favorite foods and total strangers recounted moments of my life. One woman at a kielbasa stand had already heard the one about the psychic in the tent demanding ten bucks for our brief conversation about fairs, claiming it had taken "a lot more psychic energy than a reading," which only costs five.

It was through this same rumor mill that I heard the best barbecue to be had at the Three County Fair in Northampton, Massachusetts, was hosted by a woman named Betty, a travel agent who also runs a beauty salon. Betty has come to the fairgrounds for the past seven years to pitch her tent and set up the grill on which she makes her locally famous ribs, grilled chicken, and collared greens. Propped up next to her grill was a simple handpainted sign that read: *Betty's Ribs 'n' Things*. A gentle, maternal woman with a short, graying Afro and an easy smile, she spoke to me as she sliced fat green bell peppers with expert hands. Before her was a simple wooden kitchen table that was stacked high with packages of hamburger buns and mounds of plastic silverware.

"I love coming to the fair and doing this because I love people and I love food," she told me, her hands working quickly but her voice calm and easy. "My profit margin's not that high because I like to do things right. But it's not really for the money that I do it. It's just to be part of the fair."

I asked the woman what it was about a fair that made her want to be part of it year in and year out. She paused from slicing her green peppers, her kitchen knife idle for a moment in her palm, and looked at me. At last she replied, "Fairs are about family and friends coming together and having a good time. It's babies and kids. It's husbands and wives. It's three or four girlfriends out for the day together, gossiping and having fun."

Betty pointed with a slow movement of her chin, indicating a couple who was strolling past as they held hands and pushed a baby carriage. "See that?" she said. "*That's* what I'm talking about."

She shook her head. "You get everybody here—rich people, poor people, and every age of people, from babies all the way up to the old-timers. I met this one old fella over by the race-

Fair time comes to the city, too—in a big way. The 487-acre Los Angeles County Fair, Hotel and Exhibition Complex (Fairplex) boasts more than 350,000 square feet of indoor exhibit space, a race-track with a 10,000-seat grandstand, and 12 acres of carnival grounds.

track who told me he's been coming to the fair for 50 years now. He said the only reason he came back this year was so people would know he wasn't dead yet." She chuckled and gave her chickens a turn with her fork.

"People are all the same, you know what I'm saying?" the gracious woman mused as she worked her chickens on the grill. "We may look different, we may speak different languages. But really we're all just folks."

★ ★ ★

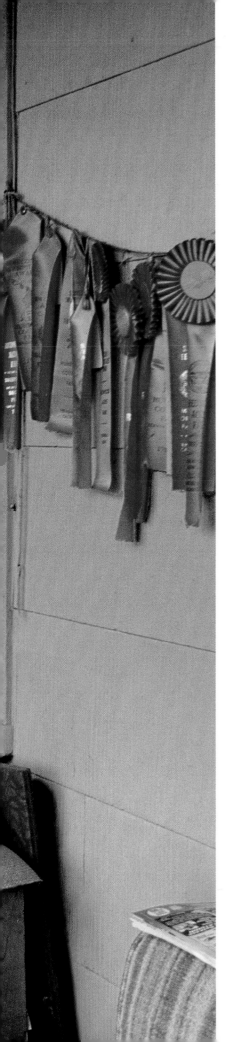

CHAPTER 2

The Exhibition Hall

At the Harris County Fair in Houston, Texas, a middle-aged woman dressed in a red sweatshirt, black polyester slacks, and glittery gold shoes moved slowly through the exhibition hall, examining a chair made out of horseshoes; a mailbox with fins and a tail; a needlepoint of The Lord's Prayer; a shag rug with Tweety Bird's face on it; a Rice Chex cake in the shape of a scarecrow; and a photograph of a dog eating a watermelon. Over the dull drone of country western music being piped into the air-conditioned hall, the woman, her hair a rigid peroxide halo, commented on each item to anyone in earshot, "Don't ya just love it? That's so *cute*! Isn't that just so pretty?"

On every fairground at every county fair stands at least one hall where the fairgoer can see the creative efforts of the community. Quilts that have taken months to create hang beside enormous pumpkins and fat ears of corn. Cakes baked painstakingly into the shape of the continental United States, or a farm or church, are displayed next to jams, soups, and fruits preserved in uniform glass jars. Just beyond them are posters

Winner of more than five dozen Humboldt County gardening prizes, 93-year-old Mary Coppini relaxes on the ribbon-swagged porch of her Ferndale, California home. "There are lots more inside. I've just never hung them up." She has retired from competition, making way for other green thumbs.

Sharp shears at hand, Mary Coppini picks her way among flower beds that brighten this northern California landscape. Mary's ribbons aren't just for flowers; she's won for vegetables, too. Fair prizes historically added welcome ready money to the sometimes cash-shy resources of farm homemakers.

drawn by local youngsters with such inspirational messages as *Baseball: America's Game* and *Always Feed Your Dog A Balanced Diet*. And dangling from almost all of these exhibits are ribbons: purple, blue, red, and white. There are ribbons for best of show, best of class, and ribbons just for participating.

Prizes (premiums) have been a part of county fairs since the first Berkshire County Fair in 1811, when sterling silver cups and tableware, for example, were awarded. The values of the premiums (not cash prizes) at this fair were quite generous: a prize valued at $10 for best bull or best Merino sheep; one for $5 each for best oxen, cows, heifers, mixed blood sheep; common sheep and swine. In 1813, the year of the third Berkshire Fair, domestic manufactures were added to the exhibitions. Competition has remained an essential theme until this day, but prizes are rarely excessive. What remains most important is the acknowledgment by your community for a task well done. At county fairs all efforts are celebrated.

Behind these carefully arranged exhibits are the exhibition hall superintendents and judges, locals who for the most part donate their time to catalog, arrange, and ultimately rank the hundreds of entries that pour in each year at fair time. For Myrna Soterakopoulos, whom I met at the presidents' hall of the Kitsap County Fair in Bremerton, Washington, the exhibition hall is about preserving traditions.

Soterakopoulos had been "slaving since a week ago Saturday" at the fairgrounds. Superintendent of the Needle Art department of the home arts display, she had been arriving at 9:30 in the morning and working until 10:00 at night, setting up and judging the entries. Assiduously going over each of the entries with a magnifying glass, she and the other judges analyzed the compositions' tension, looked for stitches crossed the wrong way, and sought to determine whether each of the pieces was executed by a single hand, or if more expert hands had interfered.

"I like to share my knowledge," she said. "And I want to encourage people to overcome their shyness, to show off all of the hard work that they've been putting into their needlework. And besides," she added with a quiet smile, "as a judge, I get to see everything first. Needlework is important. In the olden days the wives of sailors used to knit special jerseys for their menfolk so in case they drowned they could be easily identified. The kind of devotion and technique that went into things like that is being lost. I want to help preserve them. These days everybody wants to jump into the future without taking any time to look after the past. For my part, I'm not online—and I don't want to be, either."

Soterakopoulos explained that she'd been coming to the fair since she was a teenager. This, too, she said, was a matter of tradition. "If you live in Kitsap County, you *have* to come to the fair. It's the biggest entertainment of the year. My kids always spend all five days at the fair, so they all enter things so that they can get free passes. My 12-year-old son made peanut butter balls and entered them as an exhibit. He won a blue ribbon last year for those balls."

Looking around at the displays in presidents hall, I saw that other kids in Kitsap County appeared to have the same idea as her 12-year-old. Beneath a collage of heartthrobs cut from *Teen* magazine were a button collection, a pencil collection, a collection of plastic toy trucks, ten rocks in an egg carton, and 14 Barbie dolls in a cardboard box. These exhibits were displayed as prominently as quilts that had taken weeks, even months, to create. Those who live in the county can enter and can enter whatever they like. The exhibition hall is a place for us to display ourselves to our community, to show off our interests, our tastes, and our talents.

Tradition remains an important theme and traditional arts such as quilt-making and canning continue to symbolize a

county fair to many. For women like Myrna Soterakopoulos, the exhibition hall becomes a place to pass on traditional crafts and knowledge to our next generation. It provides her—and the rest of us—with a connection to our own past.

Yet like fairs themselves, the exhibition hall endures as a display case for our evolving folk ways. Seen in this way, folk art seems to be as vibrant in America today as other aspects of our popular culture. Folk art, too, changes with time. Americana, like America itself, constantly recasts itself. Nothing is sacred because nothing is static—not even tradition. In the exhibition hall of the county fair, for example, a patch quilt can hang beside a shag rug with Mickey Mouse's face on it, and both pieces are held in equal esteem.

When Elkanah Watson conceived of the county fair, he thought of an occasion that would encourage the people of a young republic around the patriotic cause of self-determination. When he encouraged women to contend in the domestic arts of canning and homespun, it was intended to achieve economic self-sufficiency within American households, and to instill a sense of pride in our own Americaness. It is not necessary for us to make our own quilts and can our own foods any longer. But the need to reaffirm a sense of our own value and skills remains as pressing today as it must have been in the early 19th century.

Fairs—like quilts and canning—may have outlived their economic usefulness. County fairs have been able to survive as economic entities largely because they are run by nonprofit groups that rely heavily upon volunteers for labor. Why do people across the country continue to donate their time to county fairs? They do so partly out of a sense of civic duty. But they also volunteer because it lends them a feeling of self-worth. As one old gentleman who was helping out at the Clay County Fair in Spencer, Iowa, told me, "It just makes you proud to be a part of something so good."

Top left *Anna Vander Kooi has festooned her Worthington, Minnesota bedroom with Nobles County Fair ribbons won in bread, clothing, and dairy food categories. The county also gave Anna her first dairy princess title, sending her on to regional, and finally statewide, competition.*
Bottom left *Sitting in a revolving, glass-walled, refrigerated room, each of Minnesota's regional dairy princesses has her likeness carved in an 85-pound block of butter. Anna first saw the royal sculptures when she was ten. "Those butter heads inspired me to try to become a princess, and I did!"*

Cecelia Hauschild would, I think, agree.

Like Myrna Soterakopoulos, Cecelia Hauschild, superintendent of the youth exhibit hall at the Three County Fair in Northampton, Massachusetts, believes that the exhibition hall is a way to preserve the past. "Canning, quilting, knitting—these are all things that I grew up with. These are our traditions. It's nice to pass them along to the kids."

Walking with me through the neatly organized exhibit at the fair's youth building, Hauschild told me, "It takes a day to get all of the entries, a day to set up, and then a day to do all of the judging. This year we had about 1,300 items to set up—and you have to make sure you get it all in! You don't want to have a disappointed kid! This morning I had a little girl come in to see if her cookies had won. You should've seen her face when she saw that red ribbon! What does a ribbon cost to get a kid going? If just one kid says 'Thanks,' then I'm happy."

Hauschild expressed to me that every year she looks forward to the fair. "This means everything to me—the time away from home, just me, doing something useful."

Although county fair competitions may be lighthearted and fun, exhibitors still take them very seriously. Chester Freeman, senior Lions Club member and one of the organizers of the

Cullman County Fair in Cullman, Alabama, told me that every year, an hour after the judging has begun, phone calls start coming in from people wondering if they've won. "Now we've got a computer—but that's still not fast enough for them," he said, smiling.

Alabama was named after the Alibamu, an agriculture-based Amerindian people who inhabited the area in pre-Columbian times. Today, Alabama has approximately nine million acres of farmland—and Cullman County has more farms than any other county in the state. Not surprisingly, Cullman's focus is farming. As Freeman explained to me, "The Cullman County Fair is an old-fashioned, bona fide agricultural fair. And for this reason we restrict participation in the fair just to people from Cullman County. We're tops in sweet potatoes and in cattle, too, and we're the leading poultry-producing county in the entire United States."

Cullman was founded in 1873 by Col. Johann Gottfried Cullmann, a German refugee who controlled land extending from Decatur, in the north of the state, all the way down to Montgomery, with the dream of creating a colony of German immigrants. When Colonel Cullmann died in 1895, he had succeeded in luring more than 10,000 immigrants to the United States, most of whom settled in the area. Today, the descendants of these German settlers represent the vast majority of the people living in this dry county of the Bible Belt. The main social event remains the fair—and possibly the world's only beerless Oktoberfest.

Just outside the cow palace at the Cullman County Fair, I met Brian Kress, 33, and his son Travis, 8. Brian was holding a basket of sweet potatoes, his son a basket of Irish potatoes. Brian, a tall, thin man with a slow, gentle way of speaking and moving, reported that on his farm he has 70 acres of sweet potatoes and 20 acres of Irish potatoes. "That's all I know,

growin' potatoes," he said. "It's what my Daddy put me to doin', so it's what I'm puttin' my boy to doin', too." I asked Travis if he liked growing potatoes. "No!" he cried. His father's long face slowly opened into a smile. "He says that because it's always the hottest day of the year when you got to pick 'em. I never much liked it when I was his age, neither."

Little Travis may not like digging potatoes, but he certainly likes showing them off at the fair. "Look at what I won!" he cried, proudly showing off a blue ribbon. "We never lost a ribbon yet!" I asked Brian how long he'd been showing his potatoes at the fair. "Ever since I can remember," he drawled. "As my Daddy always used to tell me, it's somethin' to put you in competition with your neighbors. If you got somethin', and your neighbor's got somethin', here's someone to say who's best. I've never seen anyone get a bad attitude about it. The attitude folks take here is, 'Well, if I don't win this year, I'll just try a little harder next year.'"

I wandered into an exhibition barn, where Shannon Warnke had just finished the judging for the sweet potato bake-off. On this, the third year of the bake-off, Warnke had 75 people enter their sweet potato casseroles, sweet potato breads, and sweet potato desserts. Warnke was somewhat unprepared for so many entries, and had to shanghai the young man at the booth next to hers, a tall guy with a crew cut who was trying to persuade fairgoers to raise emus, into tasting them all. When I arrived, the emu guy was looking woozy. "You never told me there were going to be *that* many!" he said wearily. "Well, I didn't know!" she protested. "Last year we only had 36 entries!" Warnke, who is just out of high school, grew up on a sweet potato farm. I asked her why she'd decided to organize the bake-off.

"To promote sweet potatoes," she replied without hesitation. "This here bake-off is just open to high school kids. The idea is to get them interested in sweet potatoes because hardly nobody my age wants to eat them. The only people who do eat them are real old, and one by one, they're all dying off. So if all our sweet potato eaters are dead, what's gonna happen to the sweet potato farmer?"

I wandered into another exhibition hall and gazed upon a series of exhibits erected by local high school students cautioning teenagers not to have sex, do drugs, drink, or litter. I also surveyed a display by a local funeral home urging fairgoers to plan early for death. In this same building were lines and lines of brilliantly colored canned goods of many varieties preserved in uniform jars.

Among these prize-winning items were the canning efforts of Connie Fortner. Fortner had entered a dozen jars that year, eight of which won ribbons. "I'm real proud of my soup," she told me, holding up a jar of vegetable soup; its painter's palette of organic colors gleamed warmly in the stark light of the exhibition hall. How did she get into canning, I wondered? "I grew up watching my mama do it," she replied. "I picked up the art from her. This year me and Mama and my sister canned more than 3,000 jars. Between us we have seven freezers filled with what we made." What, I wondered, does she do with seven freezers full of goods? "Well, we're a big family, and all of us like to eat. It's easy to go to my freezers and get out something."

Fortner informed me that she has a great many ribbons for her prolific canning. So many, in fact, that she has a closet at home filled with ribbons hanging three-deep. "I like to have the ribbons," she explained, "because I'm proud of my canning. The jars and all look so pretty, and I've worked so hard to get them just right, I like to show them off. The ribbons are just kind of a thank you, I guess."

Louise Williamson, a small, middle-aged woman, who had won 21 ribbons that year alone, is similarly enthusiastic about entering—and winning. "You name it and I do it: crafts, cook-

Following pages *In 1855 an upstate New York newspaper declared "baby shows" to be "too disgusting for serious remark." But McKenzie Gammon's five-year-old eyes (center) sparkle brighter than her sequins with dress-up and show-off excitement at Cullman County's 1996 fair. The debate continues.*

ies, canned goods, candy, everything under the sun," Williamson said, her gaze as intense as a pair of high beams. "I make 'em all and I enter 'em all. I win a ribbon for almost everything. I entered 12 cans of preserves this year and 12 won ribbons. In 1994 and 1995, I was voted Queen of the Kitchen. I had to enter at least ten different baked goods in five different divisions, and I won 'em all!" I asked Williamson what it was about these fair competitions that she liked so much. "The ribbons!" she cried. "I collect ribbons and put them up in my home to show off to people. It doesn't have to be first place. It can be first, second, or third—as long as it's a ribbon!"

Exhibitionism at a county fair is not confined to the four walls of a hall. When locals come to a county fair it is not always to compete for ribbons and prizes. Some come and display their interests, tastes, and beliefs to their neighbors. Many dress themselves up in costumes meant to communicate who and what they are—whether as a farmer or cowboy or stock car racing aficionado.

In a country that likes nothing better than to reduce ideas to slogans, the most common self-advertisement is the T-shirt. Strolling through a fair crowd, it is possible to see—or read— the myriad opinions and points of view of fairgoers. Lunching at the same picnic table or sitting side by side in the audience of the arena, are kids wearing *Dole for President* T-shirts and kids wearing *Che Guavara* T-shirts. *Rage Against the Machine* sidles up next to *Jesus Is the Standard. 460 Years of Chicano History* mills with *Cowboy Foreplay: Get In the Trunk.*

Many come to exhibit their opinions, interests, or hobbies, and to educate. Just inside the entrance to the Clay County Fair in Spencer, Iowa, stood an aesthetic semicircle of more than a hundred John Deere tractors, glinting yellow and green in the fading afternoon light.

At the tip of this enormous horseshoe sat Edward Morisch and his son, Michael, members of "one of the largest two-cylinder clubs in the Midwest." The Morisches had stationed themselves before a table on top of which was a black-and-white photograph, obviously quite old, of a man mounted atop a 1939 John Deere, along with a bumper sticker that read: *John Deere Tractors Make My Heart Go Putt-Putt*. Edward Morisch, a solidly built man with green suspenders, a pair of tinted glasses, and a green-and-yellow John Deere cap, informed me that the exhibit contained tractors dating back to the 1920s, including rare models worth as much as $30,000. I inquired how he had gotten involved in the Northwest Iowa Two-Cylinder Club.

"Memories," he replied. "A few years back I saw an old B-John Deere advertised in the paper and I recalled that back in the 1940s my Dad traded in his pair of horses for his first tractor. It was a B-John Deere."

Michael, his son, added, "A lot of people did that, traded in their horses for B-John Deeres because they worked better with horse-drawn implements than other tractors around at the time. And maybe because of that people are very close to their John Deeres. Just like people and horses, there's that deep kind of a connection."

I asked the Morisches if they took their club's tractors to any other fairs besides the Clay County Fair. "Nope," Michael replied. "This is the only fair we do because this is the only fair in the area worth doing. I'd say what we have here is the premier agricultural fair in the whole country."

Top right *Before the audience arrives for Cullman County's pint-size pageant, a shorts-clad Alabama girl ducks behind speakers to make her fairy-tale transformation. Gowned and curled, four-year-old Elizabeth Harbin will carry home a savings bond and the pride of being Little Miss Cullman.*

Bottom right *The Minnesota State Fair's outgoing Princess Kay of the Milky Way, Kimberly Mallory, has a more-than-sash-deep connection to the state's 600,000 dairy cows. Only daughters whose parents operate dairy farms, or work on them, are eligible to become county, then regional, dairy princesses.*

After spending five days in Spencer, I have to conclude that it would be nearly impossible to find anyone in that town in the heart of the Corn Belt who would disagree. Just like farmers and their John Deeres, there's a deep kind of connection between the people of Clay County and their fair. As Jack Jordan, who astoundingly had been to the fair every year since it opened in 1917, told me as he grilled up German sausages at the senior Kiwanis food stand, "State fair's a dirty word around here."

Although the population of Spencer is only about 12,000, the fair draws more than 300,000 visitors. Once a year, rising from the endless flatness of the Iowa countryside, a crowd forms—to stroll, to hear big country music acts like the Statler Brothers, to sell a grand champion boar or bull, to buy a new silo or mounted side delivery rake.

In recent years, however, the fair has served another purpose: to keep everyone together. Since the farming crisis of the 1980s, when many banks overextended credit to farmers and then had to foreclose on those not able to pay back their loans, the fair has given the farmers something they can all be proud of. One fairgoer cut to the heart of it when he said, "You know, out on the East and West Coasts they get all these new ideas that ain't new at all. 'It takes a village?' Hell, we've known that in Iowa all along."

Sometimes victory brings both thrill and agony. At the Curry County Fair in Oregon, a tearful Sarah Timeus gets a consoling hug after selling her 4-H champion lamb. Her friend Blair Jones sold his steer for $1.40 per pound and with a final, gentle ear rub said a long, sad goodbye to the creature he'd raised.

The Clay County fairgrounds even look like a village, with handsome brick buildings, carefully trimmed lawns, and uniform street signs. Within one of these lofty structures, Mary Christensen, who lives on a nearby farm, stood behind a display case of cookies and pies; behind her was a wall of glass jars—more than 700 of them—containing beans, corn, peas, carrots, potatoes, peaches, pears, apricots, and plums. While her husband volunteered his time at the cattle barn and her grandchildren showed their hogs and cattle at the 4-H barn, Christensen volunteered her time at the exhibition hall.

A very old woman was peering into the case of baked goods, her nose against the glass, intently examining a cake baked to resemble a loaf of Wonder Bread in its wrapper. She looked up at Christensen, and said, somewhat testily, "Back in 1930, they used to sell the cakes after they were judged," and shuffled off.

Christensen seemed unfazed. Surveying the wall of glass jars, I mused, "Jarring and canning must be difficult arts."

"Not really," she replied. "I do it myself. Don't enter though."

I asked Christensen how long she'd been coming to the fair.

"Forty-five years."

And how long she'd been volunteering?

"Fifteen," she replied in her conversational style.

Did she enjoy working at the fair? There was a long pause.

"We all come back," she said. "I guess something must bite ya."

At a loss for words, I stared down into the glass case and looked at a cake baked in the shape of the State of Iowa. Written in icing were the words: *Happy Birthday Iowa 150 Years.* I looked up. Preparing myself for a telegraphic reply, I asked, "Why do you volunteer?"

Christensen looked at me for a long moment. "We're proud of our fair," she said. "This is a good home fair with a friendly atmosphere. We want to keep it going. People here start talking 'fair' months in advance. They have to, to get everything organized in time. The fair kind of keeps everyone together. The 80s were hard on farmers. That sense of community we've always been so proud of here in Iowa suffered a little bit because of it. The fair forces us to stick together."

Christensen's words remained with me. Occasionally I would hear them, as spoken in her soft, Midwestern intonation, as I roamed fairgrounds across the country. Why has the county fair, that homely American tradition, survived? Why should these celebrations of agrarian life remain so vibrant in our postagrarian society? Perhaps, it occurred to me, because they have less to do with agriculture than with community, whether agrarian or not. As America becomes more diverse, and in some ways more divided, the county fairs that have survived are not those that have remained the most agricultural, but those that have succeeded in being a celebration of community—no matter how that community may define itself.

Just as agriculture has remained a leitmotif for the county fair, so has tradition. What I found, however, is that the most agriculturally valid areas did not necessarily have the most successful fairs, nor did those fairs with the most tradition. This struck me as especially true when I returned to my hometown fair in the birthplace of the county fair, western Massachusetts.

A market town with limited agriculture, Northampton, Massachussetts, is situated in a valley suspended between town and gown—and, it would seem, between the past and the future. A small city with a lot of the upscale amenities of a big metropolis, the town's business people have valiantly opposed the "Walmart-ization" of their downtown. Young people, most of them students from the five nearby colleges, dress themselves up in the street fashions of downtown Manhattan and drift in and out of the ethnic restaurants, espresso bars, cybercafés, and pricey boutiques that thrive along Main Street. Very few are

Dennis Blalock has won so many ribbons at Alabama's Cullman County Fair that his triumphs adorn him like the proverbial coat of many colors.

Three blue ribbons show that Connie Fortner mastered both the culinary creativity and food safety-precision that judges require of canning competitors.

An embroidered "Winter Silhouette" brought a second place ribbon to Sister Emilie Schmitt and cool respite to the eyes of September fairgoers.

Her gaze direct and proud, Hilary Bice held her speckled lamb close to her heart. Her brother Cole's black-faced lamb took a ribbon, too.

All the Smiths packed into the portrait as Wilma "Skeeter" Smith showed off prize-winning family produce. Sweet potatoes are big business in Cullman.

If they gave a prize for smiles, Louise Williamson could add another blue to her first place ribbons for canned pears and a "Fish stories told here" bonnet.

Gregg Hodges and James Hutchinson grew a pumpkin of prize-worthy girth and weight. Pumpkins of ideal color and conformation also win premiums.

Curtis and Opal Batemon produced outstanding field corn. Unlike midsummer's favorite sweet corn, field corn commonly goes to feed livestock.

For most of us, "pig," "hog," and "porker" are just synonyms for swine. For champ Janie Crumbley, each applies to a specific stage in the life of Sus scrofa.

It's a three-way tie for sparkle among the sleek black calf, the brilliant purple ribbon, and the broad grin that winning put on Michael Graveman's face.

Quilts are judged on design, choice of materials and colors, and fineness of execution. Artena Hunt's needlework met the judges' highest standards.

Making your hay while the sun shines won't guarantee you a tight, well-formed, dry, rot-free bale like Jason Herfurth made, but it sure will help.

aware that the fairgrounds—where the Three County Fair, one of the oldest continuous county fairs in the United States, has been operating since 1818—is only a 15-minute walk away.

For years, the biggest draw at the fair had been the racetrack, but recently it is hard to get enough horses together to run the races. As soon as you pass through the gates of the fairgrounds, you come upon the racetrack. Here, knots of people pore over race forms, smoking cigarettes down to the filters, or sit on lawn chairs set up in front of TV monitors, drinking beer from big plastic cups. There appeared to be more losers than winners on this gray day. A woman in a pair of stretchy lime green pants, a scarf knotted under her chin, kicked angrily at the litter of empty plastic cups and discarded racing forms scattered across the asphalt; another, lugging an outsized shopping bag, muttered to herself, "Worse than a house of pain."

At the far end of the fairgrounds, back near the restrooms, Capt. Russell J. Myette stood at attention. Dressed in Union blue with a red stripe running down his pant leg to identify him as an artilleryman, Myette posed next to a replica of 19th-century cannon, arms folded squarely across his barrel chest. Noticing me pass by, he called out, "Identify yourself!" Obediently, I stopped and told him who I was.

Next to his cannon, Myette had pitched a canvas tent beneath which was a carefully arranged display of pistols, rifles, caps, and shoes—all dating back to the Civil War, and all belonging to Myette. Born and raised in Northampton, Myette worked for 38 years as a maintenance supervisor at the University of Massachusetts in Amherst. After retiring in 1995, he devoted himself full-time to his living-history project. Taking his cannon and his artifacts all over the state, and beyond, he visits schools and teaches young people about the Ninth Massachusetts Light Artillery, which lost 3 of its 4 officers, 6 out of 7 sergeants, and 28 enlisted men while making a

Left *In Tillamook County's main exhibit hall, award-winning flowers fill a simple display table with bright color. At this Oregon fair, entry rules state that any article exhibited must be produced in the county. Judges consider form, color, and fragrance as they evaluate each bloom.*

valiant stand at Gettysburg. "Thanks to the heroism of the Ninth Massachusetts," Myette recounted for me, "the Confederate charge was delayed long enough for the Union line to be reformed."

A couple of grade school kids wandered over and surveyed Myette's display. He showed them a bayonet that also served as a candle holder. "With this," Myette told the children, "a soldier could write letters home while he was camped at night. Then he'd pin the letter on the back of his coat, so that if he died in battle the next day, a fellow soldier who might come across his corpse could pluck the letter from him and make sure it got delivered safely." The children, a boy and a girl, stared in wonder at the sinister-looking instrument. Myette placed the bayonet back on a table and showed them some manacles, along with a spoon that was sharpened on one edge so it could also serve as a knife. Then he asked, "Do either of you know who won the Civil War?" The girl, who was older, raised her hand as if in a classroom. "The North!" she cried.

Myette shook his head slowly; the girl's arm dropped to her side. "That may be what they taught you at school, but that ain't the truth. The truth of it is, nobody won. Who could win a war between brothers? The Civil War was about Americans killing Americans. If the two of you remember anything, remember this: The Civil War was a horrible, horrible thing to have happen to our country."

Looking as if they'd just been scolded, the children wandered away, toward the animal barns. Myette, his hands on his hips, watched their tiny figures shrink into the distance. "You know," he said, "I have no education. I worked for a living all my life. So one of the biggest honors for me was last year when I got invited to set up my living-history display over at Mt. Hermon Academy. Now, Mt. Hermon is one of the finest prep schools in the land, and they were studying American history over there; when they got to the Civil War, they called me up and said they wanted me to come over and tell them all about it. It wasn't just Americans either; there were foreign students there, too, from Europe and other places. It was something, I gotta tell you. Everyone seemed to be really interested in my display—even the foreign kids. Afterward, the teachers themselves came up and asked me, 'How did you get all of that knowledge?' They were just sitting back and listening. I can't tell you how good that felt, to have those educators come up and congratulate me. They even sent me a letter, asking me to come back next year. I have that letter at home."

Myette has invested $55,000 in his living-history display; the barrel for his cannon alone cost $8,000. It takes three trailers to haul the exhibit from place to place, and four and a half hours to set up the entire encampment. Normally he gets nothing in return.

At the Three County Fair in Northampton, he had gotten even less than usual. "I don't know what's happened to us," Myette said, his voice sinking. "Our sense of community seems to be gone. This fair here should be a reason for us all to come together, to teach one another things. But all anybody cares about are those horse races. It seems that if they lose the races, then they'll have nothing. We need new blood. The old-timers just don't care anymore. I see people I've known since I was a kid walk by and not even bother to look at my exhibit."

If Northampton was a community adrift from its own history, Los Angeles was a community with its sights fixed on the future. For me, the best example of a county fair's ability to reinvent itself was found in Los Angeles. Within a mall-like landscape of ATM machines and monumental cans of Coca-Cola, jugglers performed to the beat of salsa music and smooth-talking salesman sold everything from Jacuzzis to Japanese gardens. This was a fair as diverse and large and dynamic as the community it was set up to represent. Yet for some fairgoers, the Los Angeles County Fair had evolved so fast that it had left them wondering if some of the more traditional conventions of the county fair had any place at all.

The Los Angeles County Fair grew from a commercial-industrial show first held along the Southern Pacific Railroad siding in downtown Pomona in 1921. Inspired by its success, local businessmen launched the Los Angeles County Fair the following year. In 1996, more than a million people visited the fair, making it the largest county fair in the United States. With 8 hangar-size pavilions, a grandstand, a major horse-racing facility, 12 acres of carnival grounds, even an 8-story hotel, the fair has remained faithful to its commercial origins.

Before the Second World War, dairies and orchards flourished in Los Angeles County. Now, amid the vast urban sprawl, only a few farmers with tiny, specialized plots remain. Yet the fair survives—not so much as an agricultural event as a buoyant celebration of the diversity of Los Angeles County. To stroll through the Fairplex at Pomona is to stroll through Los Angeles, if such a thing were possible. Admiring the Mission architecture, feeling the sun, I saw a woman on stilts lope past, and a man riding a unicycle zip by in the other direction. I watched as a woman dressed as a bunch of California grapes did an impromptu dance while a mariachi band played. Someone handed me a flyer that read: *Don't let negative karma and past life experiences affect you in your present incarnation.* I examined some oils and incense being sold by a Rasta with

The home economists and agricultural extension agents who judge canning competitions may be the most serious folks on the fairgrounds. Whether the jar holds pecans, peppers, or pickles, they generally disqualify "fancy packs" that indicate too much handling for bruise-free, germ-free food safety.

Clustered on a makeshift backstage platform, Alabama high school cheerleaders prepare for their routine. Parking spaces at the fairgrounds fill up fast when it's time for the Cullman County cheerleading competition. Only the Miss Cullman and Little Miss Cullman pageants draw a bigger crowd.

impressively long dreads, contemplated buying a toe ring, and thought about getting my picture taken astride a Harley-Davidson. "Only five bucks for a memory—what a deal!" the biker babe with the Polaroids and the tattoos said.

Inside the fine arts building, three nuns dressed in white habits were watching a blond woman in a gauzy get-up do a belly dance to some Turkish music. As I watched a guy demonstrate a Sushi Master in the exhibits building, a young man with a blond goatee and a bodybuilder's physique handed me a list of all of the farmers' markets in southern California. The young man, whose name was Cary Harris, had come to the fair to help advertise the markets. "We're a group of small-scale farmers with two to ten acres of land at the most," he said. "The only place we sell is at the farmers' markets. Without them, we couldn't survive."

Harris showed me some bottles of extra virgin olive oil and some jars of honey made by his dad, a beekeeper, who had the idea of setting up this booth at the fair almost two decades ago. "Our produce isn't always cheaper," he said. "But it's much healthier, much fresher. We follow all the rules of the fair, but we use no chemicals, no pesticides to grow our crops. Everything is completely organic."

Next to us, the Sushi Master man was droning away. "You know, I've been coming to this fair ever since I can remember," Harris continued. "This building used to be the one place where you could go where everything wasn't 'As seen on TV.' Now it seems they're just trying to cram as much stuff in here as possible. I mean, come on, how many Sushi Masters and Magic Pens do you need to buy?"

The fair hasn't always been like this? "Well, yeah, it's always been very commercial," Harris said. "But you used to also be able to go and see what your community was all about, to look at displays, to get information. A county fair should be a place where neighbors can come together and relax, have some fun. Now it seems they're just trying to make as much money as they can. I mean, come on, does the California Lottery really have to have a whole pavilion to themselves? Give me a break!"

Harris told me that his father had considered not coming back this year for exactly this reason, but that he had persuaded him to persevere. I asked him why.

"Because I love the fair," he said.

★ ★ ★

Above *No longer mere adjuncts to "real" sports, cheerleaders are serious athletes, many with years of gymnastics training. Cullman cheerleading squads work hard—and loudly—to "WIN" at the fair.*
Right *In quiet counterpoint, a young boy with a livestock trophy enjoys a moment of solitary pride.*

CHAPTER 3

The Animal Barn

At the swine barn of the Kitsap County Fair in Bremerton, Washington, on a brilliantly sunny summer day in August, more than a dozen Future Farmers of America and 4-H'ers struggled, one by one, to get their lambs, steers, and hogs to walk in a circle. Looming over them, an auctioneer, perched atop a platform constructed from bales of hay, was selling the animals by the pound.

A hog had just sold for top dollar to Safeway when Matt Muzzy, a skinny 12-year-old with cropped blond hair, yanked his lamb, Butthead, into the arena.

"Now, who's going to help this young man out?" the auctioneer asked from his lofty perch. "He needs the money so he can go to the National Boy Scout Jamboree in Virginia next summer and get a badge. This is a nice-looking Suffolk here, weighing in at 119 pounds. Come on, folks, who's going to start the bidding?"

And the auction began. Speaking at 78 rpm, the auctioneer launched into an alliterative litany of numbers. His voice grew tenser, his breath

A cow's as good as a couch if you're waiting for the cattle showmanship competition at Minnesota's Nobles County Fair. For Chris Tiesler and his bovine partner to win a ribbon, Chris must manage his carefully groomed animal with poise and respond quickly to the judge's instructions.

shorter, as the numbers pushed upward. And then his voice hesitated, briefly, at $6.50 per pound. The auctioneer scanned the small group of people crowded around until, finally, from the corner of his eye, he saw a hand furtively rise from the back of some bleachers. "Sold!" the auctioneer cried.

I caught up with Matt outside the sheep barn. He was breathless; his cheeks were flushed. The throaty protests of the animals inside the barn were joined by the sounds of the crowd drifting over from the midway. The occasional breeze rolling off Puget Sound carried with it the edgy, sweet smell of fair food all jumbled up with the odors of the barnyard. Announcements somersaulted over the PA system: "Come to the rodeo at eight! See the hypnotist's act at four! Arctic Ice please go over to the curly fries stand!" I asked Matt if he was happy with the way things had turned out. He smiled, revealing a glinting semicircle of orthodontia.

"You bet!" he said. "Last year I only got $3.11 a pound."

But wasn't he going to be a little sad to see his lamb go?

Matt rolled his eyes. "It's kinda hard to get attached to a lamb called Butthead," he said, and, giving me an "Are you done?" look, sprinted off toward the lights and music of the midway.

Nothing is more emblematic of a county fair than the animal barn. It is here where young 4-H'ers like Matt Muzzy and his sister Misty can make a little extra money so Matt, for instance, can go to a Boy Scout Jamboree and win a badge. And it is here where professional livestock farmers go to compete, display their breeding stock, and get to know others in their business in a friendlier, lighter atmosphere than the one they might find at bigger, more serious-minded competitions.

The animal barns of the county fair are where kids and professionals alike keep and groom their animals, getting them ready for the hectic morning schedule of competitions that take place throughout fair week. There are competitions for

Preceding pages *On the McKenzie ranch in Langlois, bringing in the sheep can require as much hunting as herding. Thriving on the good pasture found along Oregon's southern coast, the 4-H animals auctioned at the Curry County Fair are important additions to many local flocks.* **Right** *A canny—if diminutive—stockman, John Swenson coaxes his lamb and calf toward a livestock trailer. Six-year-old John isn't a member of 4-H or Future Farmers of America yet, but he can join other Curry County youngsters who take their proudest efforts to the fair.*

cattle, hogs, chickens, dogs, horses, and even, occasionally, llamas. Next to the barn, or sometimes in it, is a ring or paddock, where animal owners compete for ribbons and money.

Typical competitions would be like those I saw at the Cullman County Fair in Cullman, Alabama. A judge, dressed in a shirt and tie and a pair of chinos, stalked about the rich-smelling ring, his hands clasped behind his back, examining a row of heifers. Their owners—young boys and girls, teenagers and adults, too—led the animals around, using long aluminum show sticks to direct them. When they wanted the animals to stop, they raised the show stick, which the animals mistook for a fence. Then the owners would scratch the young cows' bellies with the glinting stick's hooked tip, which seemed to be enough to keep the animals happy while the judge scrutinized them.

Some bleachers had been set up, and people climbed up and down to watch for a while on their way from the exhibition halls to the midway; others circled the fence of the arena, watching for a round or two before drifting off into the small but churning fair crowd. With the exception of a small clutch of friends and relatives, very few fairgoers appeared to be paying much attention to what was going on. For them, the competitions seemed to be a brief entertainment before moving on to other diversions, like the magician's or hypnotist's acts.

The livestock barns at Vermont's Tunbridge Fair have no running water. A lively procession of dairy cattle, including Jerseys, Brown Swiss, Holsteins, Guernseys, Ayrshires and shorthorns, draws welcome refreshment from a branch of the White River flowing beside the fairgrounds.

After he had watched the animals walk forward and backward and all around him, the judge, who had been taking mental notes, awarded the ribbons, explaining his decisions with compulsory good nature. I watched as the judge rated two bulls that had been paraded, a little grudgingly it seemed to me, around the ring for him once the heifers had been dismissed. After viewing them, he took a microphone in his hand and declared, "I think both these bulls have got a tremendous amount of good in them. I think they'll both go out and get the job done...."

Though professionals do show at county fairs, it seems that it is the kids who most affect the average American fairgoer. As one woman, who had come to the Harris County Fair in Houston to work in the parking lots, expressed to me, "This fair is all about the kids. Just this morning I saw a 12-year-old over in the animal barn with a turkey bigger than herself...." The woman paused to clear her throat. "It brought tears to my eyes," she continued, her voice swollen with emotion.

Dick Atkins, executive director of the Houston Farm and Ranch Club, where the Harris County Fair takes place each October, leaned back in his creaky leather chair planted in a cluttered corner of the fair office, and said, "Our biggest problem is our next generation. Fifty percent of kids today don't have two parents; they don't have a daddy to tell 'em they love 'em. You look at most of those fellows sitting in the penitentiary, and I'll bet you nine out of ten of them had a rough childhood."

Atkins, a retired police officer who is fond of quoting preachers, said that, for this reason, proceeds from the Harris County Fair go primarily to sponsoring youth activities. Peering at me through a pall of cigar smoke, he said, "This fair is set up to help the kids. When someone bids on an animal at the fair auction, it's not because they can't get the animal anywhere else. It's to help out the young people. This fair is basi-cally here to give our kids some direction, as a learning process for kids who dream of moving up some day to the Houston Livestock Show. Now, that's the Cadillac of all livestock shows. That's where the real pros are."

Denise Pulpan, who was working just outside the animal barn of the Harris County Fair selling show supplies from a trailer, laughed and said, "If you can't show an animal by now, it's too late. Showmanship is 99 percent of selling an animal. Everyone wants the top hole, the #1 slot." She jerked her chin in the direction of the barn and informed me that her son was inside grooming a steer for a competition that morning. "When your kid's got an animal to show, it forces you to spend time with him. You live with him locked up in a truck or a motel room from Friday to Sunday, and you can't help but develop a relationship. Ninety percent of the kids in that barn are good kids. You're never gonna have to get 'em out of jail for being drunk, I'll guarantee you that."

In the course of our conversation, I learned that Pulpan had grown up on a rice farm just 30 miles from the fairgrounds. "When I was a kid, this was all country. Now there are citified parts of Houston I've never even been to—and don't want to, either." Dick Atkins concurred. "At one time Harris County produced more cattle than any other county in the United States. Then the city grew and gobbled up the land. The ranchers moved out," he said.

According to the Bureau of Census, the American farm population peaked at about 32 million in 1910. When the first reports of farm population were published in 1940, about 23 percent of Americans were farm residents. By 1992, that figure had fallen to under 2 percent.

Though America may be becoming increasingly more urban, Americans' interest in organizations such as 4-H does not appear to be diminishing. 4-H clubs, symbolized by a four-leaf clover with an "H" on each leaf signifying Head, Heart, Hands, and Health, were first organized among American rural youth in the first decade of the 20th century. Funded by federal, state, and local governments, these voluntary associations, which are open to youngsters between the ages of 5 and 19, specialize in projects not related just to agriculture, but a range of interests whether it be raising lambs or studying the environment or computers or photography. Today, there are roughly 71,000 4-H clubs in the United States with an average membership of about 20 each. Only a small portion—about 12 percent—of the 4-H'ers actually live on farms.

There appears to be nothing more heartwarming for the average American fairgoer than to see a young 4-H'er with his or her animal. Undeniably, bright-faced kids and farm animals are a winningly sentimental combination. But why should the culture of farming retain such importance in our postmodern society? Perhaps because it evokes an elemental aspect of the culture upon which we've learned our country was founded, a culture of moral virtue, hard work, independence, and discipline. As America grows further and further away from its agrarian origins, the animal barns of the county fair have become an arena in which to glimpse, even capture, the values of the American past.

The connection is not entirely subliminal. A veteran carnival worker, who had moved with her carnie husband up to the Pacific Northwest from her native California because of the increasing gang violence on some fairgrounds there, told me, "At every stop, me and my husband always make a point of going over to the animal barns and seeing those 4-H kids with their pigs and whatnot. When you see so many kids messed up with drugs and gangs, it really makes you happy to see those kids with their animals because it kind of gives you hope that they'll stay out of trouble."

One kid who appears to be staying out of trouble quite successfully is Kevin Provost, a 13-year-old ox-teamster from Bernardston, Massachusetts. I first glimpsed Kevin as he stood in the middle of a hoof-pocked arena, cast in shadow by an enormous American flag that was suspended from the ceiling of a hangarlike building on the edge of the fairgrounds in Northampton, Massachusetts. Beside him were two oxen, named Kenny and Mike, which, at the judge's signal, he drove through a course marked by bright orange highway pylons. Dressed in a farmer's uniform of blue jeans, a faded red T-shirt, and a blue cap pulled down to his eyebrows, Kevin cajoled the hulking beasts into dragging a 100-pound stone boat in and around the pylons with affectionate but firm cries of "Hah!" and "Gee!" and "Whoa!" Kevin quickly completed the course, and the crowd, thinly scattered across some bleachers, clapped, a hollow, echoing sound.

I went up to Kevin, where he stood at the edge of the arena with his oxen, talking to a couple of his buddies, who were also dressed like farmers in blue jeans and T-shirts. The boys joked and laughed and manfully spat on the ground as the oxen, still side by side in their yoke, issued an occasional guttural groan. I asked Kevin how long he'd had his oxen.

"Only a couple of years," he said. "Because it was only a couple of years ago that I moved to a farm. I grew up in town, but my Mom got remarried and we moved." He patted the sweaty neck of one of his team. "These guys I named after my new stepbrothers, Kenny and Mike."

I asked if he got nervous showing in front of a crowd. "Nah," he replied. "Not at a little fair like this one, when it's just your Mom and some friends in the stands. The past two summers I've been showing every weekend, and sometimes you go to some really big competitions with hundreds of people looking at you. That's when you get a little nervous."

Preparing for the Nobles County 4-H Dairy Show,
Anna Vander Kooi grooms a heifer in her family's
southwestern Minnesota barn. "Showing animals
gives good publicity to the dairy industry," says the
ten-year 4-H veteran. She entered three animals in
the competition and brought home three ribbons.

On her way to the livestock enclosure with a heifer on a short lead (heifers are female cattle under three years old that have not yet produced a calf), Vermonter Anne Burke enjoys the relaxed assurance that comes from 30 years of experience showing animals at the Tunbridge Fair.

Loren, and my granddaughter, Lora, and pass the time. Milking's a 365-day-a-year job. You got to be there. A lot of young people don't want to be tied down. They don't want to work that hard. I guess I can understand."

Just opposite the cattle barn, at the swine barn, Kim Gacke, a fourth-generation hog man, told me much the same story. A very youthful 40, Gacke looks more like a surfer than a farmer, but his great-grandfather bought the farm in the 1890s where he and his family now live. Gacke had come to the fair to sell breeding stock, just as his father and grandfather had done since the fair first opened. But Gacke's farm has dwindled in that time from 160 acres down to the 9 that he now owns. It's so hard to compete with the new megafarmers who are now emerging in the Midwest, that Gacke said he reluctantly urged his 20-year-old boy to join the Marines rather than stay on the farm. He himself has an eight-hour job at a creamery and his wife is a registered nurse. "We're a three-job family," Gacke stated. "There's just no other way to survive."

Although a decidedly friendly event, a good deal of serious-minded business takes place at the Clay County Fair. Within the large buildings, the fairgoer can buy car wax or dinner sets, but most are interested in products that will make their farms more productive. Standing in a long pavilion where more than 30 different seed companies competed for attention, Marc Whipkey, seed salesman, told me, "This is the farmer's fair. State fairs have become more and more urbanized, more carnival-like. Here, people come to wheel and deal in machinery, equipment, seeds."

The kind of changes in farming described by Gacke and Luitjens have fundamentally altered the way men like Whipkey have come to do business. Whipkey is a natural-born salesman. He's a good listener, a good talker, a nimble joke-teller, but, most important, he's someone who understands a

changing market. Laid out before him were glossy brochures, free pencils, and a perfect ear of corn.

"It's a whole different ball game from what it was ten years ago when you'd drive out to the farm, put a leg up on the bumper, and talk about the kids," Whipkey explained to me above the din of the commercial hall. "Beginning in the mid-80s, farmers really began looking at the bottom line. Good business sense began taking over tradition. You still have pockets of heavy livestock in these parts, but more and more you see people switching over to corn and beans if they've got the acreage to do it. Investment in livestock is pretty heavy, and maintenance of the animals is very intensive. The average Iowa farmer today is 58 years old. He wants the direct return. There's much less risk harvesting corn, so the attitude you're coming across these days is, you know, 'Let's sell the corn, Ma, and buy a Winnebago.'"

Behind the commercial building of the Clay County Fair lie acres and acres of lawn, which at fair time become crowded with salesmen selling enormous, otherworldly pieces of farm equipment to successful businessmen/farmers for as much as $100,000 a piece. Lyle McCool, who represents a company that builds grain-hauling equipment, stood dwarfed by the auger carts—huge electric-green contraptions with long, giraffe-like necks riding on monster truck wheels—that he had come to deliver to a local farm equipment dealer.

McCool expressed to me, not surprisingly, that times had been tough in his business for several years after the farming crisis, when many farmers went bankrupt buying big farm equipment that they couldn't afford. "But now," he said, "over the past couple of years, the prices farmers have been getting for their crops have been getting good again. Livestock prices haven't come up like grain prices, mind you. But the farmers who've been growing the corn and the soybeans, the ones

Nine-year-old Tim Hein gets duct-taped to a pole
in the cow barn; not far away, Lloyd Winter's
daughter and her friends duct-tape Dad to his
chair. Silliness rises to an art form as the serious
business of competitions and auctions winds down
at the end of the Nobles County, Minnesota, fair.

In a shady corner of a 4-H food stand, visitors to Minnesota's Nobles County Fair arrange themselves in a grinning five-boy stack. Community groups and commercial concessionaires encourage fairgoers of all ages to indulge a year's worth of longing for cotton candy, caramel apples, and corn dogs.

Penniman's voice had brightened, but his eyes, half-hidden by the glasses he wore, remained incongruously dim. I watched his eyes as he stared out at the rows and rows of futuristic-looking grain equipment parked in endless, orderly rows on the grassy fields of the fairgrounds. Slowly, his eyes gathered light, and his face opened into a broad smile. Coming toward us I saw a small, towheaded girl, no more than three years old, holding a red balloon. When she saw her grandfather, she broke into an awkward run. Penniman picked up the little girl and gave her a kiss on her cheek, which was sticky and pink from the cotton candy she had been eating.

"It makes me real happy to see her having such a good time," he said, speaking to me but looking into the miniature, contented face of his granddaughter, Abby. "We brought her last year to the fair, but she was too young to really appreciate it. Now she's old enough to know what a county fair is all about."

The image of a grandfather holding his laughing granddaughter in his arms is a portrait of family that would make many fairgoers nod and smile and say, "Now *that's* what a county fair is all about." If the county fair is a place to exhibit our beliefs as a nation, then the one value that is exalted above all others is "family."

What is family if not a unity based on love and respect? It is a feeling of togetherness—even when times are rough. Nowhere have I seen as close-knit a family, in the greater sense of the word, than among the farmers of the Midwest. These are trying times for many of them, yet they continue to come together each year at county fairs like the one in Spencer, Iowa, to do their business, and to be there for one another. Just as Mary Christensen told me in front of the high glass wall of preserves in the exhibition hall at Iowa's Clay County Fair, "I wouldn't worry too much about the farmers of this country. They stick together."

Right *Charged with nabbing a wary rooster bound for the fair, six-year-old Jeffrey Knox reinvented a classic trap: He laid a rooster-size loop of slip-knotted string on the barn floor and baited it with feed. Result? One handsome bird in hand, and one Curry County, Oregon boy beaming with pride.*

America's a very different place from what it was in Elkanah Watson's time. Farming, like everything else in our society, has undergone fundamental changes. Yet farming—and farmers—continue to represent an aspect of our national identity that we seem very much to want to preserve.

I cannot help but think again and again of Emma Hohenberger. Despite the gloomy times some professional livestock owners have seen over the past decade, Emma seems determined to center her life around livestock. For this young woman, her animals are not just a hobby—a way to have some fun, and maybe pick up a little pocket money at fair time—they are her future.

Watching her vigorously brush her shorthorns, I asked Emma her plans for the future and, without hesitation, she replied, "My aim is to get into a major agricultural university and study international livestock management. I'd like to work with different cultures and their animals. And I'd like to help develop a market for American beef overseas. That's my dream."

Emma looked around at the other cattlemen busy working their heifers and at the schoolkids with the ribbons they'd won tucked into the back pockets of their jeans. She watched exhibitors, both young and old, maneuvering their large charges into the arena for the next competition, and said, "It's a dream that was born right here, at little county fairs just like this one."

★ ★ ★

CHAPTER 4

Entertainment

In conservative New England, where the county fair was born in 1811, entertainment was often considered frivolous. But in its beginnings as an agricultural fair, "rational amusement"—activities that educate and entertain—was part of its makeup. As the county fair evolved, wholesome competition was found on the fairgrounds. These competitions were specifically related to farm activities, which were meant to honor farming prowess, such as plow matches. These activities also included informal nonfarm competitions, such as foot-races and wrestling matches. Although somewhat serious-minded, the competitions were a kind of entertainment for local fairgoers.

County fairs became the social event of the rural year, and entertainment began to be performed for the sake of entertainment. At first locals provided this entertainment, whether by participating in competitions or by joining in parades or by singing agricultural hymns. Soon professional entertainers started appearing at fairs, providing diversion for a price. For many rural areas, fair week came to serve as a kind of Sears

Transformed into a makeshift dressing room, a forestry exhibit tent bustles with baubled, bangled, and beaded members of the "Shaken Not Stirred" Middle Eastern dance ensemble. Local talent—from the familiar to the exotic—flourishes here in California's Del Norte County and at fairs across the country.

catalog of fun, delivered once a year to their community's doorstep. The entertainment tent became a place where dreams came true. Locals could re-create themselves as stars, and peripatetic professionals were able to find the audiences that they hoped would one day catapult them to fame.

Like fairs themselves, fair activities evolve with time. There is a good example of the change in county fair entertainment in plow matches. At the earliest fairs farmers competed against one another in plowing and drawing matches—who could draw the straightest furrow, for example. These competitions focused on who could best drive an animal pulling a plow. When tractors began to replace draft animals in the fields of America's farmland, the animals were also usurped in the drawing competitions of county fairs. As American society developed, and its tastes along with it, tractor pulls would be performed alongside pick-up pulls: flashy, hyped-up events where semiprofessionals compete beneath the glare of bright spotlights.

At the Clay County Fair in Spencer, Iowa, I saw just how far plowing matches had come since the early 19th century. As a yellow hot air balloon wafted across the cloudless Iowa sky, a thick, black column of smoke rose from the grandstand, where earlier that afternoon, Frankie Valli and the Four Seasons had performed. Wandering inside to the edge of the arena, I peered through a high chain-link fence. I watched as John Deere tractors, with tall vertical exhaust pipes, dragged great weights over a distance of several yards. Commanding the tractors, which not only produced a great deal of black smoke but also made a spleen-rattling amount of noise, were competitors from all over the Midwest. They were dressed like race car drivers, with white helmets and white stripes shooting down the lengths of their red leather outfits.

After the tractors a battalion of pick-up trucks roared into the arena. They were dragging what looked like the flatbed of

Top left *Hoist a 35-pound wriggling pig. With your free arm, crank-start a stripped-down Model T Ford. Jump behind the wheel. Clutch, shift, accelerate, brake, and steer around the track. (Don't lose your pig!) Make three laps. Stop and restart your Ford three times. Carry three different pigs. Is this madness? No, it's a Pig-N-Ford race!*
Bottom left *Parry Hurliman (and his porcine passenger) brace for take-off in the 68th running of the Pig-N-Ford races at Oregon's Tillamook County Fair. The winner gets a trophy and his picture on the front page of the Tillamook* Headlight-Herald.

a 16-wheeler loaded down with a tremendous weight over a short but laborious distance. "Hoo boy!" the announcer's huge, disembodied voice echoed. "Look at that! Man, oh, man! That's just a little ol' Ford engine settin' in there, folks!" For almost two hours the show continued, as Fords competed against Chevrolets. Providing a narrative to all of the smoke and noise, the announcer called out, "Pretty neat, wouldn't you say, folks! Hey! Are you Ford people still out there? You're in the lead! You Chevrolet people, your time's acomin'! Now here, folks, is a 648 cube called Nut 'n' Honey all the way up from the Show Me state, Mizura! Let's see what she can do!"

While big spectacles such as the pick-up pull draw large audiences, simpler entertainment continues to find its corner on the fairgrounds. Just in back of the fair office at the Kitsap County Fair in Bremerton, Washington, beneath the shade of some leafy trees, I watched as a preschool-age girl in a frilly white party dress confidently climbed onto a small stage. After taking a microphone in her hand, she proceeded to belt out an extremely high-pitched, but seemingly very heartfelt, rendition of "Stand By Your Man." Her song over, the little girl took a deep bow and, in no uncertain terms, commanded the thinly scattered audience to come up to the stage so she could

curtain that sparkled like tinsel, he interviewed the contestants. The little girls were all between the ages of three and five, but they were emulating teenagers on prom night, their hair stiff with spray, their little faces all made up. A row of judges in the front row took careful notes, rating the girls on a 70-point scale: 30 for poise, 30 for personality, 10 for dress.

After the contest, a small, extremely affable gentleman approached me. He was wearing a glow-in-the-headlights vest, and on the back was a big heart drawn around the state of Alabama with the words *Cullman: Heart of Dixie* written in the center. A senior Lions Club member and one of the organizers of the fair, he told me, with an urgent sense of pride, "That Little Miss contest you just saw is exactly what our fair is all about—families. I think you might like to know that of the 25 little girls who entered that pageant, only one was being raised by a single mom."

As communities evolve, so do their fair queens, and sometimes faster than the communities they've been elected to represent. At the Harris County Fair in Houston, Texas, I ran into Kaylyn Dickey, 17 years old, a pretty and gregarious blonde. I identified her by her sash and crown as she wandered around the enormous barn where LeAnn Rimes, "Thirteen-year-old country western singing sensation," was gearing up to play for a beery evening crowd. Some high-school-age cowboys had stopped to get Kaylyn's autograph, which she signed with a flourish as she made each boy turn around so she could use his back as a writing surface.

I introduced myself to Kaylyn and her mother, Hope. "The cowboys are just a little extra bonus that comes along with the crown," Kaylyn remarked as she waved good-bye to her new admirers, who disappeared into an outsized paddock where a few people had now gathered, practicing the two-step in the dry, powdery dirt.

Top right *Among the delights of fairgoing are the spectacles not on the program. Dressed more for the Grand Ole Opry than for the shed-roof pool hall at the Brazoria County Fair, a young Texas boy warbles into a headset microphone as his parents play Nashville hits on a portable stereo.*
Bottom right *Fair time offers precious opportunities for adults to cheer as youngsters succeed. Clarence Woods heads up a line of proud escorts as he waits to accompany his soon-to-be stepdaughter, Katelyn Steward, in a procession of Junior Fair Court Princesses at Oregon's Curry County Fair.*

Kaylyn touched the rhinestone crown that sat atop her feathered hair and said, "When I won this, I said to my mama, 'Look! Now I got a crown of my own! I don't have to try on my friends' crowns in their rooms and just pretend anymore!'" What did it mean to be a fair queen? Kaylyn thought for a moment, then said, "It means to be a good role model. It means to be nice. Those are my duties as fair queen." I asked her what made her enter the pageant. "Well," Kaylyn said, "I'd always wanted to be in a pageant, but I never dreamed I'd actually win one! I was competing against a whole mess of real pretty girls—and they were all real cowgirls, too. They weren't just dressing up like me."

I glanced again at mother and daughter, who were both dressed quite convincingly as cowgirls in tight jeans, cowgirl boots, and western-style blouses. They weren't the real McCoy? Hope Dickey took a ladylike swig from a can of beer, and confessed, "We're from the other part of Houston, the southeast part, with the gangs and the drive-by shootings and the SWAT teams and all. Out here, this is real country. This is the way Houston used to be."

Kaylyn nodded. "We live in both worlds. Don't get me wrong, country's fantastic. In fact, back in the eighth grade I

The Traveling Choir from Texas' Ramsey III prison puts on a pop-rock-gospel show at the Brazoria County Fair. Contact between family members and inmates at public performances is not permitted, but this woman exchanged a quick hug with her husband before guards intervened.

used to be something of a cowgirl myself. I used to get a whole bunch of compliments on my kicker jeans, too. And even now, if some friends want to go to the rodeo I'll be there, the first in line, having a good old time. But this is 1996. There's the urban part to Houston, too. Like tomorrow, me and my Mom are going to this street festival over in the gay district with alternative bands and people wearing nose rings and all. It's a real lot of fun; we go every year. That's where you see everybody in Houston—blacks and Latinos and whites—all together, whoopin' it up."

For many professional entertainers, a stop at a county fair is just one leg of a much longer journey. The journey is a literal one, leading them all around the United States and Canada as they follow the tight, hectic schedule of fairs that runs throughout the late summer and fall. And for others, like the magicians Gifford and Roy, it's just one step toward a much bigger dream.

At the Three County Fair in Northampton, Massachusetts, situated at the end of a long gauntlet of fast-talking carnies, stood the Coca-Cola tent. It was a green-and-white-striped canopy, beneath which was a stage concealed by a black backdrop with the words *Razzle Dazzle '96* written in glitter letters. The sun began to set, and the crowd started to thicken as the lights of the carnival rides beckoned against a blue-pink sky. Fairgoers beneath the nearby beer tent struggled to match the volume of the polka band performing for them. Just then an announcement ripped through the fairgrounds: "In just a few short minutes, right here at the Coca-Cola tent, before your very eyes, the mystifying illusionists Gifford and Roy will command the stage and engage the imagination as they lure you into their world of magic and wonder! You will be amazed and mesmerized at what is appearing and disappearing before you! Experience the magic! Witness their illusions! Step right up, folks! The show begins in just a few minutes!"

And begin it did. In a cloud of dry ice and a riot of disco lights and a burst of dance music, Lance Gifford and Robert Roy, dressed in matching black slacks and black satin jackets, leapt onstage into a bit of well-staged choreography. As the music switched with a DJ's seamless precision, from Enigma to the theme from *Mission: Impossible* to "Macarena," Gifford and Roy capered from one illusion to the other. Roy disappeared into a "Hindu Basket," pierced with swords, and, miraculously emerging intact, was put inside a box that was cut into threes, from which he also materialized in one piece. The show progressed as Gifford swallowed razor blades, pulled doves out of thin air, turning the birds into a rabbit then a poodle. Then they reminded the audience that the illusions of Gifford and Roy could be seen during the whole month of October at Spooky World in Berlin, Massachusetts. Finally, the two invited fairgoers to come up on stage and have a souvenir picture taken with a python around their necks.

Later, inside their trailer parked behind the Coca-Cola tent, Gifford, dressed in a white polo shirt and a pair of blue jeans, was drying his hair with a towel. He sat between a computer and a caged parrot that squawked "Hello!" to anyone who came inside. Gifford, 28, told me that he got his first magic set when he was 12 and, inspired by the world of magic and illusion, joined the circus at the age of 16. The circus ran out of money three weeks later at the fairgrounds here in Northampton. Undeterred, Gifford pursued his dream, teaming up with Roy, 25, nine years ago.

Traveling up and down the east coast, Gifford and Roy perform mostly at fairs, both county and state, and at an occasional convention. "It's a challenge performing at big events like this one," Gifford said, "because there are so many distractions. You have to draw people in and hold on to them. But it's not just at fairs that it's hard. We just worked this science fiction fair in Boston. The cast of 'Star Trek' was there. It's tough holding on to people when Scotty's out there signing autographs."

This is not their only challenge. Unlike the suitcase magician, illusionists like Gifford and Roy have an enormous amount of stuff to travel with. It takes two trailers to haul their stage, their equipment, and their impressive menagerie—which includes pythons, a parrot, several doves, rabbits, and dogs—from fair to fair. As well as being quite unwieldy, all of this gear is very expensive.

"So far we've invested $200,000 in our act," Gifford said. "The box with the blades that Roy goes into and uses to cut paper into thirds with and then swallows before your very eyes? That cost $4,200. It took two months to make—we had to measure Rob specially for it. We've had the stage we're working with now for about seven years, but now we're looking to get a really super-duper one, but it's going to cost about $90,000. But everything we have we've worked for. We do three shows a day, and we travel most of the year, from New England to Florida. It's expensive, and the animals like to eat a lot. But we want to keep growing, so at this stage everything has to go back into the show. We've never seen anything even close to a profit yet, but we're patient."

And they are hard-working. When Gifford and Roy weren't performing their act three times a day, they were running across the fairgrounds to oversee the pig races. Dressed in matching T-shirts with cartoon images of pigs on them, Gifford and Roy bustled four hogs out of a hot pink trailer decorated with big black polka-dots and an American flag suspended from its bow and the words *Granny's Racing Pigs* across its side. As "Old MacDonald" played scratchily, the two magicians shuttled the snorting beasts into the starting gate. With the blow of a whistle they were released and, somewhat half-heartedly, pursued a short, circuitous course. It was defined by

Following pages *Brazoria County Fair manager Anita Rogers confesses that the midway "probably brings more people through the gate than any other single attraction." But Texans will still turn out for dashing riders, whether they are rodeo professionals or equestrian showmanship competitors like these.*

a very low chain-link fence that was marked by black-and-white-checkered racing flags. A crowd of amused fairgoers gathered around to watch as Gifford provided a sportscaster-like voice-over to the animals' reluctant progress. One pig, whose name, Gifford informed the crowd, was Tammy Faye Bacon, suddenly lost interest in sniffing noisily at the ground. She waddled through the course to its end, exciting cheers and laughter from the crowd. As the audience dispersed, wandering next door to the Furry Friends Zoo, or to the racetrack, or the midway, I approached Gifford and Roy. They were attempting to give out piggy souvenirs to some lingering fairgoers.

Gifford and Roy bundled the swine back into the pink polka-dotted trailer. "I hate pigs," Gifford confessed as he wiped his sweaty forehead with the back of his arm. "But it's nice at the end of the season when we donate them to a 4-H barn. But what can we do? We need the money for our new stage. Try going to the bank for a loan when you're a magician. When they ask you what you do for a living, they say, 'Forget it!'"

In the off-season, which is usually two or three months in the winter, Gifford and Roy test gas. "You have two identical cars. One car has Shell in the tank, the other has Exxon," Gifford explained. "They travel the same 15,000-mile route." Did he enjoy doing that? "Not especially," Gifford said. "It's actually pretty boring. But it gives you time to make strategies for the future. The illusionist business is a tough one. You've got 28 working magicians in Las Vegas alone. But if you work at it, you can get there. Eventually we'd like to grow out of the

county fair circuit and start playing bigger venues—auditoriums, that sort of thing." He smiled. "That's our dream," he said, and then, apologizing, ran off to the other end of the fairgrounds for their next magic act.

At the Los Angeles County Fair, fairgoers go for entertainment in a very big way. Walking through the orderly, mall-like fairgrounds, the fairgoer is confronted by many sights. There are mariachi bands and rock bands and square dancers and belly dancers. Chinese acrobats fly weightlessly through the air and Olympic swimmers dive from dizzying heights. Elephants also offer lifts and computer terminals are available for rides on the Internet. Even the more modest, homespun entertainment has a glitzy edge to it in Los Angeles.

There is not just one entertainment tent at the Fairplex in Pomona, California, as almost every structure on the premises is used for some sort of amusement. Even the building where 4-H'ers had hung their hand-drawn posters to patriotism and common sense was used. Here, I saw a troupe of young people, who called themselves "Kids On Stage For A Better World," perform a series of songs and dance routines accompanied by a synthesizer. I watched as a young girl, maybe about 14, interpreted a song about the environment, written by L. Ron Hubbard. A dozen others sang back-up while an eight-year-old in a flowing ballet skirt did a few interpretive ballet steps in the foreground.

After the show I talked to some of the performers, who explained that their group was an outreach program of the Church of Scientology. Performing at malls, schools, parks, and festivals throughout southern California, the kids try to transmit an anti-drug, as well as an anti-gang, message to their peers. "We can reach a lot of people here at the fair," said Matt Bartilson, 15, a dancer and actor who occasionally appears in sitcoms. Jierra Clark, another young member of the troupe,

nodded in agreement. "We've had about 900 people pass through this building just today. We can't touch all of them, but at least some of them will think about our message." And what is their message? "Our message is that you can make your own dreams come true." And what are her dreams? Clark smiled, winningly. "My dream is to be a professional performer," she said.

Magic, illusion, dreams—these, it would seem to me, are the themes that animate the entertainment tent of the county fair. The entertainment tent, itself, varies in form—sometimes it is a tent, sometimes it is a proper auditorium. Although its structure might vary, the spirit that occupies it is very much the same.

The county fair has become increasingly entertainment-oriented since its first conception in the early 19th century, just as America has done. Though fun has become the ruling motif of the county fair, this fun would seem to serve a bigger purpose than mere amusement. In a nation where entertainment has become an enormous, exportable industry, the homely entertainment of the county fair continues to thrive because it appears to encourage the smaller dreams, hopes, and beliefs of everyday Americans.

On no occasion was this more clear to me than at the rodeo. On a weekend in late August I sat at the top of the grandstand at the Kitsap County fairgrounds, looking out at a Ferris wheel turning lazily against the bluish Olympic Mountains that loomed behind it. Echoing throughout the fairgrounds was the much-bigger-than-life voice of John Wayne praising the beauties of America. When the recording ended, the rodeo fans, who were squeezed together all around me, started stomping their cowboy boots, causing the aluminum bleachers to shudder. Then the rodeo announcer declared, "Welcome to Pepsi-Cola day here at the rodeo!" Then a cowboy, gripping the mane of a bucking bronco, shot out of a gate, followed by two

Hats full of donated dollars put smiles on the faces of Del Norte County dancing girls. Some of that money likely came from lily farming. Set between giant redwoods and the Pacific Ocean, this northern California county grows 90 percent of the lily bulbs commercially produced in the United States.

girls in iridescent sequined shirts, one carrying the American flag, the other a flag with the Pepsi emblem emblazoned on it. I watched, my heart pounding, as cowboys wrestled calves, rode broncos, and were flung from the backs of bulls.

After a while I gave up my seat high in the bleachers and wandered below to the cramped, noisy pens. Here, the broncos and bulls remained corralled, waiting impatiently to bolt into the arena and toss a brave cowboy to the dirt. This is where I met Troy Ryan, a 28-year-old born in Saskatchewan, who had just ridden a bull for a roaring crowd. It was a relentlessly hot day, as if the normally moody weather of the Pacific Northwest had chosen to cheer up just for fair week. Ryan's face was shiny with sweat. His large-brimmed cowboy hat cast a shadow over his eyes, but his clean-shaven jaw shone brightly in the sunlight. Although it was impossible for me to read his eyes, I could sense the adrenaline still rushing through his frame.

"This is the biggest thrill there is," Ryan said, his chest heaving, his breath short. "You can't imagine the excitement when you first shoot out there into the ring. Absolutely nothing can beat it." Ryan then informed me that he only started riding bulls when he was 23. This is relatively late for rodeo cowboys, many of whom start roping calves when they're still in grade school. Despite his late start—and a cracked eye socket he received three weeks earlier when a bull threw him—Ryan was determined to ride his dream as far as he possibly could. "This is just a warm-up for me," he said with a grin. "Right now I'm just testing the waters for the pros."

After the performance, I wandered among the stables and trailers that lay huddled within the long semicircle of shadow cast by the arena. I ran into Don Kish, rodeo livestock man. Blond and tanned and square-jawed, Kish glared at me through a cloud of sun-shot dust and declared, "Rodeo is the

last true sport." Kish, who provided the animals used in the rodeo, told me that although he had been raising animals since he was 12, he had never really planned on becoming a rodeo livestock man. It just happened.

Kish explained to me, "I really wanted to be a ball player, where the big money is. Then the baseball strike happened, with all those belly-aching millionaires, and I just got sick. Now I don't even watch the games on TV. It was then that I realized that the rodeo cowboy is the real star. There's no one there to pick him up if he gets thrown. He doesn't have an agent—it's just him and the telephone. The best cowboy there is won't make more than $150,000 a year. And even if he's made it to the top in this business, he's still out there, feeding hay, driving his own pick-up from show to show, sleepin' in the barn if he has to."

I asked Kish why he had decided to bring his rodeo animals to the Kitsap County Fair. "You still got good rodeo in Washington. The life still exists up here. And at the fair, you got people who appreciate entertainment—good, family entertainment. For five bucks, you get a good show. The working man can't afford nowadays to go see basketball and baseball, what with expensive tickets, and on top of that your overpriced beers and hot dogs." He paused, his hands on his hips, and looked around. "This here is the real world," he said.

★ ★ ★

A radiantly relaxed smile replaces pageant poise as
Melony Stricklin, the reigning Miss Cullman
County Fair, brushes through a sparkling curtain.
Young women from eight Alabama high schools
competed for her title; Misty Michelle Gardner of
Holly Pond High School took home the tiara.

Above *Comedian Phyllis Diller prepares to perform in Pomona. With more than a million visitors, the L.A. County Fair is larger than some state fairs.* **Following pages** *Sideshow trickster Angelo Helay swallows swords and becomes a human pincushion when he isn't breathing fire across the fairgrounds.*

CHAPTER 5

The Midway

A day at the fair may begin with livestock, but the night belongs to the carnival. As daylight drains from the sky, the lights of the carnival beckon from the midway. This is when teenage friends and couples make their way to the honky-tonk lights and loud pop music of the carnival— to roam, to see, to be seen, to bravely use four-letter words. It is the flipside to all of the wholesomeness and nostalgia of the fair. The carnival is a rite of passage and a place to find adventure. This is when the carnies, those 20th-century American nomads, take over as masters of ceremonies with their guttural cries of "C'mon, buddy, give her a try!" and "Hey, pal, step right up, you look like a winner!"

Carnivals were first organized in the early days of Christendom as a three-day celebration preceding the long and penitent season of Lent. By the Middle Ages these celebrations had become enormously popular. Jugglers, clowns, and acrobats came to perform for the crowds of merrymakers, who feasted on ready-made food provided by street vendors eager to make a little money off of all the conviviality.

A high-wheel cyclist in period costume pedals through the Canfield Fair in Mahoning County, trying to draw visitors toward an exhibit on northeastern Ohio's industrial history. But with carnival lights flashing and the smell of popcorn wafting through the air the midway's spell is hard to break.

In the United States, Phineas T. Barnum first transformed the amusement show into big business. Dubbed "the Prince of Humbug," Barnum understood that the "public loves to be fooled." Traveling the country with the legendary midget, General Tom Thumb, and the original Siamese Twins, Barnum promoted his itinerant show with great pomp and not a little exaggeration. His original show, which eventually grew into Ringling Bros. and Barnum & Bailey Circus, created a whole new industry in the United States—that of the traveling carnival. By the turn of the century carnivals had become a fixture in both county and state fairs. Today, just as in the past, the most prominent feature of any carnival at a county fair is the rides. The most traditional is the merry-go-round. Merry-go-rounds derive from a tournament entertainment, popular among European aristocrats of the 17th century, called a *carrousel* in French or a *carosello* in Italian. A plebeian version of the carousel was invented by a Parisian toy maker who placed hobby horses on a circular platform that was manually rotated. Though carousels would eventually be propelled by motors, they continued to carry toy ponies dressed in fancy tournament dress.

In the United States, the carousel reinvented itself in vertical form as the Ferris wheel. The first Ferris wheel was built for the World's Colombian Exposition in Chicago in 1893. An enormous, almost overwhelming construction, 250 feet in diameter and able to carry 2,160 riders, the Ferris wheel was a symbol of America's bold self-confidence.

As America grew and prospered, the Spartan ethics of our farmer-ancestors came to coexist with a brazen, entertainment-oriented culture. It was an overwhelmingly popular culture, and the carnival was a sublimely base expression of it. By the turn of the 20th century, both freak shows and girlie revues were traveling the county fair circuit, where, for a small admission price, the unexposed rural American could glimpse fat ladies and strip shows and what came to be known as "glomming geeks." These last were professional trenchermen who would eat live snakes and chickens for a horrified, though fascinated, audience. While deformed babies and burlesque shows have lost their appeal for most fairgoers, the carnival as a whole retains a kind of perverse, yet beguiling charm.

If the fairgrounds is like a village, then the midway is the village green. Its spinning, flashing Ferris wheel inevitably provides the fairground with its most prominent profile. Like a secular church spire, it rises above all the other fairground structures, both permanent and makeshift, that huddle around it. That glittering, rotating wheel is the orientation point, whether you are in the animal barn, the exhibition hall, the entertainment tent, or the commercial building. Like a beacon, the Ferris wheel draws all fairgoers to the midway with its promise of adventure, escape, and chance. Today, carnivals continue to be inextricably linked with fairs because they seem to provide a predictable, but necessary, form of liberation for the communities they visit. On the midway, one can comfortably flirt with danger within the security of family and community.

Despite the very modern garishness of the rides and lights, the midway retains a distinctively medieval feel, although it long ago lost its Christian context. A colorful, hectic bazaar that creates itself once a year in an otherwise empty lot or pasture, the carnival transports the fairgoer to another time and place. With its narrow lanes lined with loud and aggressive hawkers, the midway forms a kind of 15th-century village set up within the sprawl of the American landscape. Americans can lose themselves here in a small world where all the senses are assailed at once. Wandering through the narrow lanes, tripping over the thick cables that snake through the grass, the fairgoer can smell and gorge on greasy-sweet carnival food; look upon the bright, flashing lights of rides with names such

Carnival workers—carnies—bring the midway to life with their games and rides. At Oregon's Curry County Fair, as on midways everywhere, grease is a fact of life for these nomads. The black gunk smears hands and clothes but it also keeps the mechanical swoops and swivels of their rides safe and smooth.

Silly hats and giant snacks are midway classics. An Oregon fair volunteer sporting a jaunty moose cap waits her turn at a stand selling elephant ears. Old-fashioned pastries compete with newfangled fare such as Dippin' Dots—tiny supercooled ice cream pellets—for fairgoers' appetites.

as Chaos, Sizzler, Zipper, and Avalanche; and hear, over the epileptic shudder of the motors that drive the huge looping and spinning rides, fragments of outdated pop music playing from a dozen different sources. Here, the three high priests of the carnival reign: the spinnerhead, who runs the games; the ride jock, who oversees the rides; and the concessionaire, who serves the sort of overpriced, overly sweet food that can best be appreciated within the atmosphere of a county fair.

Food on the midway is served from what carnies call "grab joints." Typically grab joints are U-Haul-like trailers dressed up with pennants and lights and an occasional fast food icon, such as a giant two-dimensional potato, perched on top. Carnival food is a gastronomy all to itself. It includes foods like corn dogs and funnel cake, and every sort of edible item served on a stick: hot dogs-on-a-stick; corn-on-a-stick; mushrooms-on-a-stick; even cheese-on-a-stick. It is an eclectic menu, offering churros and pizza and kielbasa and stuffed potato skins. At one fair, I even saw a grab joint ambitiously advertising "a six-course sirloin dinner."

Gargantuan portions of this food is, for some, the order of a day at the fair. I recall wandering through downtown Cullman, Alabama, biding my time until the Cullman County Fair began at 5 p.m. Cullman is a pretty assemblage of brick buildings with steep green canopies shading taxidermy shops and gun/pawn shops and a store called Heaven's Helping Hand Bookstore. Taped in one front window, crowded with tasseled, overstuffed Victorian-style furniture with upholstery as thick as armor, was a sign that read: "*Ye Do Err Not Knowing the Scriptures.*" Not too far away was a café, where I stepped in to have a cup of coffee.

The place was empty except for two waitresses, who were endlessly cleaning the chrome and faux black-and-white tile with vigorous strokes of sponges and rags. The acrid smell of the cleanser they used swallowed even the faintest odor of food. As I looked over some notes, sipping my cup of coffee, I eavesdropped on the women's conversation. I ordered another cup of coffee, while the waitresses' conversation turned to the fair. The younger of the two, a spunky redhead in a pair of cut-offs and a T-shirt, who had moved behind the long Formica counter to clean its already spotless surface, announced that the night before she and her boyfriend had gone to eat at the Buena Vista restaurant, whose vista is the midway at the Cullman fairgrounds.

"I saw all kinds of rides that they never had before," she said. "I just upped and told my old man that I wanted to go to the fair right then and there, even though we'd already made plans to go tonight." As she scrubbed, a smile came to her lips. "I don't know, I just got this urge to go and ride some of those new rides and eat some of that funnel cake. You don't get that fair food but once a year and somehow it just tastes different. It has that county fair kind of taste."

The older woman shook her head, and declared, "I'm never so sure about that fair food. You don't know where it's been for all its travelin' around. You're always best to eat before goin' to the fair, that's what I always say."

At the stroke of five, the carnival at the Cullman County Fair came to life—and the fair along with it—with that year's county fair anthem, "Macarena." The carnival, which had traveled from Nashville, had a distinctly macho appeal. Strolling through the alleys of the midway, I saw a game called Ring-A-Knife, where you throw plastic rings around one of many types of daggers—sleek Swiss Army knives, kitchen knives with jagged teeth, ferocious-looking hunting knives—stabbed into a wooden table. Another stand allowed you to throw baseballs at a row of beer bottles in order to "git yerself a red dog." At the carnie's feet was a litter of broken glass, and above his head,

Young fairgoers make the most of every minute, relishing candy apples while waiting in line for a carnival ride. "It's fair to say," chuckles Clay County Fair manager James Frost, "that the first time they come to the fair most rural youngsters have never been in a more crowded and busy place."

dangling as if from hangmen's nooses, were animals with red synthetic fur. Further on, pistons hissed as people rode inside giant red cowboy boots with mean-looking spurs that churned around a pole with a big red cowboy hat on top.

Standing within the penlike confines of his game, a former truck driver and now full-time carnie, was working the crowd. "I need a winner here, folks! Step right up! It only takes a dollar to play and a dollar to win!" Helping him with his game, where you throw rings around the necks of brown, long-necked bottles and win a big stuffed animal with the word "Luv" sewn across its belly, was 17-year-old Justin Haynes. He was a local who had joined the carnival just for the "spot," meaning he would stay home when the carnival packed up and moved on at the end of fair week.

I asked Haynes why he had joined the carnival. "Ever since I was a little kid I came to this fair to play the games," he said. "I wanted to know what it was like on the other side." He turned to his boss. "How'm I doin'?" he asked. The carnie didn't answer, just told Haynes "to go snag" a passerby. Once Haynes' back was turned though, the boss lowered his voice and confided to me, "He's real good. I'll turn him into a carnie before we leave town. He's too young to join this year, but I'll get him next year."

Even though Haynes wasn't a full-time carnie, he was still treated like a member of the family. "We're like any family," the game man said of the carnival workers. "Sure, we have our fights, but we never let them get out of hand. And the most important thing is that we take care of each other. And it don't matter what color you are—black or white—we stick together." The carnie took a deep breath, filling himself with the smell of fried dough and diesel fuel. "This is the life." He sighed. "The travel, the crowds, the lights. It don't get much better than this. This is the best fair we do. It's packed from open till close."

Following pages *A glittering slide and whirling Skydiver ride lure attention from the business of poultry and sweet potatoes as the Alabama moonrise transforms the Cullman County fairgrounds. "Folks come 44 miles up from Birmingham," says fair manager Odie Johnson, "just for the rides."*

Haynes, who had succeeded in snagging two or three players, called out, "Everyone in Cullman is here. Fair time is something we all look forward to. Rest of the year the only thing to do around here is drive up and down Highway 31." He waved to a group of passersby who cheerily called out his name. "The fair is 'the thing to do,'" he said, quoting the phrase with his index fingers. "Everyone waits around for the fair to come so they can walk around the midway and run into friends."

Just as there exists a distinct culture to the exhibition hall or the animal barn, there exists a culture to the midway. For the fairgoers who head there, the culture they uncover is one of mystery, of fantasy—and of experimentation. In the present, just as in the past, the midway is a place of firsts for many people. For some, like the two 13-year-old best friends whom I met in Northampton, this was the first year they had been allowed to come to the fair unchaperoned. Making the most of the time before their designated rendezvous at sunset with their parents, the boys proudly told me that upon passing through the fair's gates at noon, they had gone straight to the midway. Once there they had eaten as much junk food as they possibly could and ridden every single ride.

As the sun set and the two boys were safe—and perhaps more queasy than sound—with their parents, older adolescents roamed in packs through the souklike passageways of the midway. In the atmosphere of chance, generated by the neon lights and pop music and deceptively easy games, rites of passage were performed, whether they were a daring ride or a

"I'm an overachiever in silly contests," laughs Robbie Elmore. Twirling 27 hoops simultaneously Robbie won her sixh L.A. County Fair hula hoop competition. From butter churning and hog calling to ice cream eating and sheep shearing, fairgoers enter (or watch) 88 contests over the fair's 24 days.

sneaked cigarette or, perhaps, a stolen kiss from a mysterious someone from another school district.

For the carnies themselves, the magicians who create this fantasy world out of thin air, the midway is a way of life. Singletons and dreamers, iconoclasts and wanderers, they zigzag across the United States for more than half the year. They visit towns, both small and big, offering everyday Americans the chance—even if for only a night—to escape. Coming together from all walks of life, the carnies have developed over time a subculture of their own, including a vocabulary comprehensible only to them.

Just as in the time of Phineas T. Barnum, the carnival remains something of a house of mirrors. In the spirit of illusion, perfected by the Prince of Humbug, carnies have not always portrayed things as they are. In the argot of the carnival then, customers become "marks," and "sticks" are often employed to carry extravagant prizes up and down the midway with ecstatically victorious expressions plastered on their faces. On almost every midway are games of skill, called "grind-stores," many of which are more difficult to win than they would seem. "Grinding," rather than ripping off, is what carnies like to call it.

For most carnies, however, it would seem that there is more pleasure in seeing people happy than in grinding them. Most spinnerheads earn an hourly wage of no more than five to six dollars. For them, there is little incentive to cheat a kid out of his allowance. Rather, they have joined the carnival because it provides them an arena in which to be themselves, to interact with others on their own terms. Many refer to the carnival life as "show business." They think of themselves as performers, priding themselves on their own particular style.

At the Three County Fair in Northampton, I met Sonya Anderson, 16, a local girl who had joined the carnival—but only for a week. A friendly blonde with an easy smile and a glinting nose ring, she stood behind her Flip-A-Chick stand, where you toss plastic chickens into a big black pot, and informed me, "You have to be people-oriented to be a carnie."

Anderson told me that she moved to Massachusetts from Florida when she was five—and had been coming to the Three County Fair ever since. She was always fascinated by the carnival, she said, and when she was old enough, she started chatting with the carnies. "The whole thing to me is fantastic: the travel, the people. You have to like people, and like seeing them happy. There's nothing better than seeing a little kid win something. Their little faces light up and it just makes you feel really good to see them so happy. It's all luck, that's what I like about it. Like my Flip-A-Chick game, it doesn't take any special skill; just a little bit of luck. That's what's wonderful about the carnival: Anybody can feel lucky. It's a fantasy, a place where all of your dreams can come true. Even as a little kid I would just wander around the midway, looking at all the lights and all the rides, and think to myself, 'Wow! Wouldn't it be great to just run away with the carnival?'"

Did she still fantasize about running away with carnival, I wondered? "I fantasize about it, sure, but I won't do it," Anderson replied, then gave her trim tummy a gentle pat. "I have a little one on the way," she said, smiling beatifically. "I have to think about responsibility now, not fantasy. You can't live a fantasy your whole life."

Some, though, have made fantasy their life. At the Kitsap County Fair, I met Sid Morris, veteran carnie, who helped explain some of the mysteries of the carnival. Morris, a youthful 52-year-old, was dressed in a T-shirt and shorts, a row of silver bracelets on her tanned forearm, and a line of silver hoops in her left ear. A concessionaire, she was operating—appropriately enough for the state of Washington—a movable

espresso stand. It looked like a type of movie set café, with white plastic chairs and tables and a potted geranium on top of each. At the end of fair week, it—and Morris—would pack up and move on. Morris kept an attractive stand, with pretty bottles of different colored syrups lining the edges of her gleaming chrome coffee bar. Despite the fact that the carnival would only be in town for five short days, she had already memorized the orders of her regulars—like me. At my second visit to her stand—called a "joint" in the lexicon of the carnival—she had my double espresso already primed in the machine.

As she served espresso to a long line of fraying fairgoers, Morris chatted easily with me. She told me that she began her more than two-decade-long career as a spinnerhead, which, in her case, meant she manned the toy machine guns. Since she was good-looking and a woman, she'd always get "the first loc," or the best spot on the midway. This was usually somewhere near the entrance, or next to the entertainment tent and its regular crowds. She did so well with her machine guns that she became an "agent," which meant she graduated from getting paid an hourly wage to earning a percentage of the game's profits. From agent she graduated to concessionaire.

Morris explained that there's an age-old, though good-natured, rivalry among the carnies who run the games and the carnies who run the rides and the carnies who work at concession stands like hers. "Ride jocks don't mingle with spinnerheads. It's a competition thing, for customers," she said. And the food people? "We're the spoiled brats of the lot." Is there a name for food people in the idiom of the carnival, I wondered? "If there is one, people have been too polite to repeat it to me," Morris said and laughed.

A teenage boy interrupted us to ask Morris if she needed anything. "Not right now, honey," she said, and he replied, "Okay, Mom, I'll come by a little later." The young man, who

was dressed in hip-hop style, with baggy shorts, an oversized T-shirt, and a baseball cap turned backwards on his shaved head, ran off. I asked if the young man was her son. Morris laughed. "No, no! He calls me Mom because I'm a manager, so I've got seniority. When we carnies say we're a family, we mean it. We take care of each other. This kid has had a pretty rough background. When he joined up, me and my husband took him under our wing. We bought him a tent and a sleeping bag. We watch out for him—not just me and my husband, but everyone in the carnival."

Born and raised in Los Angeles, Morris and her husband, who is also a carnie, worked in southern California for 20 years. Seven years ago they moved to the Pacific Northwest. She liked it up here "in the sticks," and she liked the fairs, the small, hometown fairs like the one in Kitsap County. "The fairs up here are pure, simple, like they used to be," she said. Did this mean that fairs changed a great deal in the more than 20 years that she had been working them? Morris replied, "Fairs haven't changed; areas have changed—like California, which is why my husband and I got out of there. Now, at some of the fairgrounds, you have drugs around, and gangs—kids who don't come for a good time, but just to look for a fight.

At some of those fairs, things are more cutthroat. It's harder to get into the family."

Later, when business slowed down and she had time to chat, Morris told me that she had two grown-up children, a son who just graduated from bartenders' school, and a daughter, who has a degree in psychology and is married to a serviceman stationed in Germany. Neither, she said, got "the bug" to join the carnival. This did not bother Morris. She was grateful that her children were grown up now, and that she had been successful in keeping them away from drugs. Morris told me that even she wasn't planning on working in a carnival her whole life. "It just got in my blood," she said.

I asked Morris what made her stay in the carnival so long. She smiled as she looked out over the sand-castle skyline of the midway rides. It was funny, she said, but just that morning she had been telling her helper, a local girl, what it was about the carnival that she loved so much. It was "slough" night, she had concluded. In the parlance of the carnival, Morris explained, slough night is the last night at any given spot, when the carnival tears down its illusory village and moves on.

"It's the adrenaline of the whole thing," she explained. "It's the sound of the aluminum hitting the ground. Just as the fair is about to close on the last day, everyone in the carnival is edgy, watching out for the lights on the Ring of Fire to go out—that's the signal. Then everybody just works as one, ripping everything down—all the rides, all the games, all the stands. It's that funny feeling, both happy and sad, knowing that one spot is finished and over with, and it's time to move on to the next."

Despite the sense of belonging that carnies find within the carnival, Morris admitted that they are sometimes regarded with caution by outsiders. "There's a bad stigma about carnival workers," she told me as she pumped a shot of espresso into

Dressed to be noticed in go-go boots and mini-dresses, young ladies at the L. A. County Fair exchange flirtatious smiles with a friendly carnie. As evening comes and the crowd grows, many a young man will try at stubborn length to win one of the giant stuffed animals for his favorite girl.

a plastic demitasse cup. "Usually, though, people are pretty nice to us; they'll keep the bars and restaurants open until after the fair closes so that we can come in. Some places are tough, especially at new spots, where the people don't know you. Then people can be a little suspicious, a little less friendly."

As America has grown, it has inevitably become more jaded. If unscrupulous carnies did take advantage of naive farmers in the past, that naïveté has long ago been replaced by a stiff wariness of these annual visitors. At the Cullman County Fair, the county agent, a tall man with short, graying hair and large, aviator-style spectacles, propped a boot on the back bumper of his pick-up. He looked out over the midway and shared an opinion that I had heard many times before, and at many different fairs, "Ninety-nine percent of the people here would rather not have the carnival here. But they know that it's what gets grandma and granddaddy to bring the kids."

The magic of the carnival was not only lost on the county agent. It was also lost on SFC Jeff Findlay, a sturdily built 32-year-old with a red crew cut. He stood a stone's throw from the midway at Northampton, beneath a green recruiting tent, which was next to a sign that read: *100% Tuition for College!* The tent was pitched in a slow spot, between the stables and the arena, where horses were drawing buggies around and around for a judge while a middle-aged woman played an electric organ. It was a limp, humid day, the kind I remember far too well growing up in Massachusetts, but beneath the recruiting tent it was dark and cool. A full-time Army National

Guard recruiter, who is supporting his wife and son while earning a bachelor's degree in business administration from Anna Maria College, Sergeant First Class Findlay explained that he was at the fair "to sell education."

While enthusiastic about the Army National Guard, he was somewhat less enthusiastic about the increasing carnivalization of America's county fairs. "I don't get fairs," he confessed to me. "I'm from Florida, that's where I grew up. There weren't any fairs around my town, so I never went to any. And there weren't any pigs around either, so I never had a reason to join 4-H. I've been living in this area now for the past eight years, and in that time this is only the second time that I've been to the fair—and this time I didn't have any choice. First time I went for the horse racing."

He peered out from his tent, at the bit of horse track that was visible through its drawn flaps, and at a few bedraggled fairgoers, who were wandering by as they gnawed on corn dogs and dragged a screaming toddler by the hand. "I think fairs are dirty," he said. "And expensive, too. Earlier I got hungry, so I went to look for some food. I bought a potato, and it was terrible, so I bought a sausage grinder, and it was terrible, so I bought a bottle of water. The whole thing cost me $8.50—and the only thing that was any good was the water."

He shook his head and said, "It'd be a cold day in hell before I brought my three-year-old son here. I wouldn't let him come now, and I won't let him come even when he's 10 or 13—no matter how many friends he has along with him. I know all about the rites of passage thing—I did it myself when I was a kid: the jumping over the fence and then running like heck…and then everything that comes after it. No way, I'd never let my boy come to the fair. And the carnies! Now, they're a scary bunch of guys. I had one recruit who suddenly stopped showing up. It turned out he'd got to talking to the carnies and decided it'd be a great idea if he joined the carnival. We got him back, though. First thing I asked him was, 'What the hell were you thinking?' He said he thought it'd be cool to see what it was like. I told him: 'You don't up and do a thing like that when you're a 20-year-old man; that's the sort of thing you think is cool when you're a 9-year-old kid who doesn't know any better. Now, put your life together and get back in the National Guard!'"

On the midway the extremes of the fairgrounds converge, bringing the innocent and the hardened face-to-face. This is where kids and carnies come together and share in a culture that is as contradictory as it is elemental. Children are inextricably linked with the midway, just as they are with the exhibition hall, the animal barn and almost every other county fair venue. Why, I wondered as I traveled from fair to fair, are kids so important to fairs and to the people who attend them? Certainly these occasions provide us (or so we hope) with a bridge between our past and our future. It is possible here we can pass on our traditions, our beliefs, and our customs to our next generation.

But I've come to suspect that there is something else to it. When I—like so many others that I met—think of a county fair, I recall the late summer days I spent as a child roaming the grassy alleyways of the midway. Perhaps a county fair reminds us of lost innocence. Perhaps when we return, year in and year out, we are hoping to remind ourselves of our own ingenuousness.

As the county agent and SFC Findlay and countless others know, innocence is a fragile thing. Naïveté would seem to be especially vulnerable on the midway, which in some ways represents the darker aspects of a fair—and of ourselves. The carnival is, after a fashion, about willingly coming of age. When we enter the precincts of the midway, we do so in the hope, conscious or not, of "getting lucky."

Midway entertainment captivates even multi-media-jaded teenagers. Boys and girls at the Brazoria County Fair enjoy a little romance, a little comedy, and a little playful revenge as "Bozo" strafes the Texas crowd with jocular insults, drumming up business for his dunk-the-clown booth.

"We do all our business at fairs," says Steve Bennett, owner of Arizona-based Photos Now. "It's very popular. People like to take home mementos." A booth-side mirror captures Susie and Rick Gomez waiting for their black-and-white portrait at the sprawling Los Angeles County Fair.

Luck provides the midway with a constant motif. Strolling among the huge, mechanical carnival rides, listening to their cicadalike drone, the fairgoer sees crude, low-tech games that, like Sonya Anderson's Flip-A-Chick game, don't require any skill, "just a bit of luck." One such grindstore requires you to toss a tennis ball into a milk can; another to pop a balloon with a dart; another to toss a Ping-Pong ball into a tiny goldfish bowl and win a fish—"*Aquarium Sold Separately*," a hand-drawn sign qualifies.

Just as America is a less innocent place than it used to be, the carnies who run these simple games of chance don't seem quite so lucky as perhaps they once did. On one midway I met a very unwell-looking gameman with a hacking cough. It was a slow afternoon and he was having trouble attracting fairgoers to his hoop game. "It's a tough life, the carnival, and it's getting tougher," he told me, balancing a scuffed-up basketball on a thin leg. "Everything's getting a lot more expensive, the cost of footage on the midway, the price of stuffed animals—everything. As a carnie, you're out on your own, with no health insurance or nothing. That's a tough way to live. A townsman'll come up to you and say, 'Hey, buddy, how come you got no teeth?' Well, who's gonna take me to a dentist? Used to be maybe someone from the town you'd meet would take you. But not anymore."

In some places, such as the Clay County Fair in Iowa, locals appear to have given up on the carnival altogether. Most farmers are already on their way to bed when the carnival lights begin to blink, unnoticed, at the far end of the animal barns. "You follow more or less the same route year after year, so you know what to expect," said Jason Magida, 49. He is a personable, energetic man, who, along with his wife, travels with a carnival based in Austin, Texas. At the Clay County Fair, Magida knew not to expect very much. "This is a farmer's fair," Magida

told me from his station on the midway, a mechanical race-track with miniature plastic horses ridden by tiny plastic jockeys. "These people don't do games."

It was nighttime in the Corn Belt, what should have been the busiest time for the carnival, but the midway was eerily empty. Besides the occasional halfhearted cry, "How 'bout you, mister? Wanna give it a try?" the midway was quiet. It was so quiet that you could make out, above the groan of the machinery, all of the words to "I Shot the Sheriff," which was being played at the faraway ride, the Himalaya. The carnies, looking zombielike at their rides and games, had, it seemed, given up as well. "A game like this needs crowds," Magida said. "But tonight I've already done three crossword puzzles—without interruption, I might add."

Magida suddenly switched on his headset microphone, the kind an operator might wear, and called out to a couple of teenage girls whom he saw drifting past, "Girls! Girls! Free game if you can tell me who was our eighth president!" The girls gave him an "I dunno" look, and continued walking. Magida shrugged and switched his microphone off. "That's a special technique of mine," he said. "I call it incidental knowledge. I don't know how I know all this stuff, I just do."

Magida told me that he wished there were people on the midway that night, not only to save him from the tedium of doing crossword puzzles, but so I could see him in action. "Ask anybody on the lot about the horse-race guy," he said. "I'm into what I do—you have to be, otherwise you're not going to draw players. I was doing the water race before this for five years, and then started this game about a year ago. I like the horse races. It takes longer, but I get to walk back and forth and b.s. And I can play around with people. I've got this little toy bat, so I can hit people playful-like over the head with it if they're messing up."

Magida joined the carnival 10 years ago after a varied career. He told me, "This is the life—the travel, the people. I've always had wanderlust; I've always loved people." He paused and scanned the empty midway. "My wife and I have a trailer. We live comfortably. And the work suits our lifestyle. I'm happy."

The carnival, it seems, can accommodate all types of lifestyles—maybe in a way that society at large cannot. Of all of the "families" that come together each year at fair time, surely the most diverse is that of the carnival workers. Within its embrace, lapsed priests and ex-cons, spiritualists and sinners, find a way to be the indomitable individuals that they are. At Bremerton, I met Gary, 23, a ride jock whom I found slumped in front of a monumental piece of machinery. It was frozen in place, its motor silent, because of momentary technical difficulties. Gary, who felt it unnecessary to give me his last name, told me that he had joined the carnival just for three spots, all here in Washington State.

In his brief experience as a ride jock, however, he had come to realize that there exists a gentle rivalry among the three categories of carnival workers. "We're considered the grease monkeys, the bad boys of the lot. But the way I see it ride jocks are the good guys. We just want people to have a good time. People really like it when we make the rides go fast. And we don't get any tips or anything, not like the food people. The most we'll get is maybe a dime or a quarter that flies out of somebody's pocket or something."

When Gary was not doing a part-time stint as a ride jock, he was a full-time tattoo artist. He says this is a sensible profession on rainy Puget Sound because you can work indoors. Covering his arms and legs were psychedelic arabesques in sea blues and greens. He told me that after the fair closed that night, he had an appointment to do a tattoo for a friend of a friend. "Something with space aliens," he said.

Gary explained to me his vision. It was his opinion that the tattoo culture was simply part of a much bigger, much more important culture that exists in the Pacific Northwest of this country. Speaking in the slow, careful speech of that region, where people take the time to pronounce every syllable, he announced that he was proud of the grunge culture that was born in and around Seattle.

"Culturally speaking, everything new and interesting that happens in America starts right here," he said. "At first people in other places think it's weird, then a couple of months later, everyone's doing it, whether it's the way they dress, or the music they listen to, or whatever. People here are more relaxed. Here, we take the time to look at the trees, to enjoy the weather." I asked Gary to define this new American culture for me. He thought for a moment, stroking his neatly trimmed goatee, and looked up into the cloudless August sky. "It's all about being the way you want to be."

Fairs—like America—have an extraordinary ability to embrace change. Just as the entertainment arena has evolved to accommodate monster truck rallies and the exhibition hall every sort of contemporary kitsch, the fairground as a whole reflects a culture in continual flux. It is a living, changing thing, a product just as much of the farmers of the Midwest as it is of the tattoo artists of the Pacific Northwest.

County fairs are places to celebrate our common culture, and they are also occasions for us to honor our subtle regional differences. From California to New York, our American landscape of freeways and malls and Taco Bells often seems to change little from place to place. But from this sameness emerges new ways to view ourselves—like the grunge culture that Gary took such pride in.

Jack Fanelli, a carnival operator, took time during the busiest part of the day to ponder the poetry of the carnival. I met

Matt Smith holds onto his hat and onto girlfriend Monica Fairbanks. The 13-year-old Texans enjoy the Brazoria County Fair, their smiles as bright as midway lights. "With the security we have," says fair manager Anita Rogers, "parents seem comfortable leaving kids on their own for the evening."

Fanelli, who along with his father, John, and his brother, Jim, run Fanelli Amusement Company, at the Three County Fair in Northhampton, Massachusetts. Peering at me through a grilled window in the side of the trailer that served as their office, Fanelli sighed impatiently when I asked if he could talk to me a little bit about the carnival life. "This is a busy time, we're just settin' up," he said, giving me an imploring look. It was late afternoon, just a couple of hours before sunset, which, in the carnival business, was prime time. Finally agreeing to a chat, he fielded my questions while snatching for an incessantly ringing phone and managing harried-looking carnies.

Fanelli told me that his father first started the business in 1956. In a season that runs from March until the end of October, the Fanellis will set up their carnival at probably six or seven county fairs. Fanelli was especially proud of their many rides, such as the Pharaoh's Fury, Qadzilla, and Wipeout. "You must've gone around and seen the top quality equipment we've got," he said to me between phone calls. "Off season we take everything back to Fitchburg, Massachusetts. We've got a maintenance facility there, where we keep everything running like silk."

I asked Fanelli what were some of the biggest headaches of running a carnival. "Some of the biggest headaches?" he asked, looking at me. "It's a lot of stress. It's long days—16-hour days, 7 days a week. But probably the biggest headache is finding qualified people and holding on to them. It's the nature of the business, the traveling, the being away from home. Single adults, adventurists—they're few and far between."

I mused that it must have been an enviable childhood, growing up in a carnival. "You'd think it'd be paradise for a kid," he replied, "but it isn't like that. I've got four children of my own, and it's hard on them sometimes—just like it was hard on me and my brother when we were little. We're apart a lot of the time, just like I was apart from my father. What's

Preceding pages *The Ferris wheel debuted at Chicago's Columbian Exposition in 1893, less than 30 years before the first Los Angeles County Fair was held in Pomona. Today's biggest, fastest, most brightly lit rides are called "spectaculars," and their drawing power can make or break the midway.*

summertime anyway? It's amusement rides, sure. But it's picnics and days at the beach, too. We don't get to do none of that. I have a new baby girl. But she's not with me right now; she's at home with my wife. And this weekend, a cousin of mine is getting married out on the Cape. Everybody's out there right now, having a good time. And me? Well, I'm here. There are sacrifices in life," he added.

Surely there were some pleasures in running a carnival? Fanelli smiled and leaned forward a bit. The phone rang again, jarringly, but this time he didn't pick it up. Instead, he pointed over my shoulder toward the midway. I looked. It was a beautiful Labor Day weekend, the best day I had seen at the Three County Fair, with the cheeriest weather and the fullest, happiest crowd.

"Sure, there are some pleasures," he said, his brown eyes softening. "Like right now. What's happening out there. It's the kids, their excitement, the thrill they get in winning something. It still touches me. And it's the people you meet, people from all walks of life. You get a real feeling of family from everybody working in the carnival—from management all the way down to the guy who picks up the garbage."

Fanelli shook his head. "You know, sometimes when you're not traveling, when you're home after the season, when it's the dead of winter, and it's snowing outside and you're shut up in your house, you miss all this, you miss a day like today."

★ ★ ★

Two friends grimace together as they face The Zipper at Oregon's Tillamook County Fair. To ride, or not to ride? That is the question. One Zipper fan recalls a young boy whose pre-ride bravado vanished almost instantly, replaced by full-throated hollering: "Mommy! Get me out of here!"

At Iowa's Clay County Fair a video camera and microphone capture the mounting tension of Ejection Seat riders while bungee cords attached to two 125-foot towers stretch toward the ground. **Right** Riders hurtling 160 feet skyward squeal in delighted terror. Heads turn on the midway.

Following pages *With the Ferris wheel twirling behind her, Susan Stober stretches out and enjoys a clear view of the nighttime audience packing the grandstand at the Tunbridge Fair. The surrounding Vermont hillsides offer cool, calm retreat from the crowded hurly-burly of the midway.*

CHAPTER 6

The Commercial Building

If the county fair is where Americans can find their values enshrined, then capitalism has its own altar in the commercial tent. To enter the Kitsap County Fair in Bremerton, Washington, is to first pass through the commercial building. This is where pitchmen aggressively sell not only Ginsu knives and hot tubs, but political ideas, and even religion. The long building is as frenetic as a bazaar, where booths compete for attention and a fistful of dollars. Their signs cry out: *Vinyl Siding! Hair and Nail Academy! Family of Freemasonry! Waterloo: The Ultimate Solution to Clogged Drains! The Bible Story! Retirement Plans! Choose Life! U.S. Army! Elect Lowans Judge! Lady Desiree Lingerie Home Show!*

After spending an entire fair season traveling to county fairs, both small and big, it seems to me that the commercial building is in many ways much like the country itself. This is not only because the desire to make money is as American as apple pie. It is also because, in the true

At the Tunbridge Fair in Vermont, Yankee ingenuity triumphs over Yankee frugality. Fairgoers swarm through the commercial tent, tempted by trinkets, gadgets, and friendly flimflam. Traveling salesfolk and local craftspeople alike rely on fairs to draw crowds—happy, curious, and ready to spend.

spirit of county fairs, the commercial building, more than any other place on the fairgrounds, is as rich, varied, and diverse as America. Here, the hopes, dreams, and ambitions of every kind of American find their most cacophonous voice.

At the commercial building, exhibition and entertainment, the two essential themes of a county fair, come together. I witnessed this firsthand as I watched Jon P. Zoffka, a master of the pitch, at work. Wearing a headset microphone, he stood somewhat elevated in a booth outside one of the exhibition halls at the Clay County Fair in Iowa. He was selling sets of kitchen knives. With clipped, breathless delivery, he told the expanding semicircle of fairgoers: "Just as you've seen on TV, folks, these miracle blades come with a lifetime guarantee. These are the best knives you will ever own. With this knife you can cut wood, cut metal; we even cut the steel of this hammer head. I even know someone who cut their way out of jail with this knife. Don't laugh! You're looking at him—just kidding. Actually I'm just out here because we're trying to raise money to put Mom through welding school. Yeah, she used to be an alligator wrestler down in Florida, but she got tired of winning—just joking, folks. No, but seriously, these are terrific knives, each and every one of them made right here in America. You can cut just about anything with them. Just this morning I had a lady come up and say she cut a cow in half with this knife. And that's no bull!"

As Zoffka went through his routine, making the onlookers chuckle, curious Iowans gathered around to see what was so funny. His pitch appeared to work. At the end of the performance, fairgoers started queuing up to examine his miracle blades firsthand. Making some sales, Zoffka announced the timing of the next demonstration and switched off his microphone headset. Even though their audience had now vanished, Zoffka and his sidekick, a man younger but just as manic as

himself, continued to kid around. They were doing Teddy Kennedy impressions for each other and wiggling the blades of their knives in the sun to cast trembling reflections on the vast, windowless wall of the exhibition hall.

"People ask me, 'How did you get to be here selling knives?' And I always say, 'A long string of bad luck,'" Zoffka told me with a laugh. Originally from Iowa, though now a resident of Arizona, he has been a pitchman for the past 20 years, selling not only knives, but all kinds of kitchen equipment. "I like it, I'm my own boss," he said. "And when I'm not traveling, I'm in Arizona, playing golf." Zoffka travels across the country to both state and local fairs. Next he was off to Massachusetts for a big regional exposition.

While the pitchman of a century ago may have had few tools other than a soapbox and a sense of humor, today the pitch has gone high-tech. Besides hawking his knives at fairs, Zoffka has also produced his own infomercial. It, he reported, was playing on Japanese television.

Despite such ventures, Zoffka still likes coming to fairs, and pitching to people directly. "Coming to fairs like this, you meet a lot of people, and 95 percent of the people you meet really are pretty nice. I like giving them a good time, making them laugh. It's nice, coming back to a fair you've been to before because you'll get people who'll stop because they remember you from last time. Afterward they'll come up and say, 'We bought a knife set last year; we just wanted to come by and watch.' Fairs like this one are nice. The people out here in Iowa are the salt of the earth. At some of the really, really big state fairs, you lose that human touch."

That human touch is what makes a county fair different from any other kind of fair. One sees a varied crowd at a county fair—and the commercial building is no exception. Pitchmen, I discovered, are as varied as what they try to sell. At

Serious agribusiness is conducted at Iowa's Clay
County Fair. In nearby fields farmers test-drive
state-of-the-art equipment from manufacturers such
as John Deere and Case IH. In a quiet corner of
the fairgrounds, a display of vintage tractors sparks
reminiscences of harvests both bountiful and lean.

the Three County Fair in Northampton, Massachusetts, I wandered along a muddy lane where a handful of pitchmen were trying to get the attention of a thin stream of passersby. I saw a woman selling some sort of 20th-century snake oil and a man demonstrating an Electrolux vacuum cleaner. Next to him was a woman selling psychic readings from a tent for five bucks a shot.

I wandered over to where a middle-aged man was selling orthotics, pieces of plastic molded to the shape of your insole to help you walk better. The man, whose name was Tom Morisette, stood before a table on top of which were plastic models of a couple of feet gone wrong—one extremely flat, the other with an alarmingly high arch. "Do you have some unhappy feet?" he called out to a woman who walked by, feeding on an enormous serving of onion rings. The woman paused and instinctively looked down at her toes. "Are you on hard surfaces a lot? What kind of shoes are you wearing?" The woman proffered a scuffed-up tennis shoe. "Hoo boy!" Morisette exclaimed. "My favorite kind of customer! Step on in and let me take a pedograph of your foot. In less than three minutes you'll be walking on air—guaranteed! Step right up, Ma'am, don't be shy! One hundred million Americans are walking better today thanks to the miracle of orthotics!"

The woman continued on in her inadequate shoes as I went to examine Morisette's display. Morisette, I discovered, was originally from Kennebunk, Maine, where, until 10 years ago, he owned a store selling insulation and other energy products. One day a man walked into the store and offered to buy it. Morisette sold it to him on the spot and moved with his wife to Florida. They were 51 years old at the time, about 20 years younger than most of their neighbors, and they soon got bored of night after night of bingo. Morisette began to scan the back pages of *USA Today* for career opportunities. This is how he

came upon the idea of selling orthotics. Now he and his wife spend their winters at Epcot Center selling orthotics to other retired people. The summers find them traveling the United States in their Airstream Trailer—a silver, bullet-shaped home on wheels—with $20,000 worth of inventory and a tent, stopping at fairs like this one. "A lot of retired people do it," Morisette told me.

A value greatly honored by Americans is entrepreneurialism—and it is an inclusive one. It is only fitting that this value should find its place on the county fairgrounds. It is a place where retired people like the Morisettes can go as they pursue their post-retirement life in their Airstream trailer. It is also where young men like Adam Almeida can embark on the first leg of a journey that they are certain will lead to fortune.

"Entrepreneurial, that's me," said Almeida, his even white teeth flashing against a face deeply tanned from a summer of picking blueberries. I met Almeida at the Three County Fair in Massachusetts, just a stone's throw from where Morisette stood with his display of plastic feet. A business student, Almeida, 20, was not wasting time. He was using his summer vacation to take night classes at a local community college and to learn, firsthand, the ways of capitalism.

"At the blueberry farm," he said, "you get paid three dollars for each 25-pound box of blueberries you pick. Most people pick about 35 boxes a day and are happy. Not me. I averaged 70 boxes a day." Now that blueberry season was over Almeida was at the fair, where he had set up his own mini-carnival. He

If you think you're not ready for a real teeny-weeny
bikini, a Los Angeles County Fair entrepreneur
will sell you a shirt guaranteed to fool the eye, at
least for one glance! The fair vendor's motto: Build
a better mousetrap—T-shirt or window squeegee—
and the world will beat a path to your booth.

Broadcasting live from the Del Norte, California, fair, a reporter interviews a Lions Club volunteer. Fair business looms large in many counties: Local organizations often rely on fair revenues for support throughout the year, and chambers of commerce promote their fairs as proof of community vitality.

showed me some of his games, which, he told me proudly, "were all invented, built, and painted by myself." There was a miniature golf course, a roulette wheel, and a game he called Plinko—a vertical board with nails hammered into it enclosed by a piece of Plexiglass into which the player dropped Ping-Pong balls. "You see," Almeida told me, eyes wide with enthusiasm, "the beauty of these games is you can never lose. No matter what happens, you win something."

He showed me some of the prizes: old comic books and stacks of baseball cards wrapped in cellophane that he called "grab packs." Almeida explained that he started collecting the cards when he was about 12. "I'd get whole collections from friends for wicked cheap because either they'd outgrown them or they just needed the 50 bucks—and that would be for a collection that was maybe worth hundreds." He picked up one of the grab packs. "Look," he said, "see what's on top? That's Roger Clemens. That's a really good card. That's what you do, you put some really good cards on the ends, like Clemens or Kirby Puckett or Will Clark—you know, stars—and in the middle you just put a bunch of junky repeats. But even these good cards, which are maybe worth 50 cents apiece, I'd've gotten 'em for a third of a penny, so I don't care; I get a buck each time someone plays a game."

Having shown me his games, he opened a cooler that held some bottles of soda floating in ice water. "See that?" he said, arching a thick, black eyebrow. "See all the different flavors? You're not going to find that kind of variety anywhere on the fairgrounds. Variety—that's key; that's how I keep my edge."

Almeida informed me that he had started his business career at the age of eight, when he got his first paper route. "I wouldn't just throw the paper over the fence and ride away on my bike," he said. "I'd stop and talk to people. That's how I started understanding people and what they wanted. Then, when I was 12, I started working at flea markets, selling stuff, and that helped me even more in developing my business sense."

And what about the fair? How was business here? "Oh, boy," he said and grinned. "The fair is fantastic—the people, for one. You can't get a better crowd than at a fair. You got everybody here. And, of course, the money—you can't beat the money. People come to the fair to spend."

Americans seem to love fads, and a county fair appears to be a great place to market those fads. Just as fads sweep the country, they storm the county fair circuit as well. In the fair season of 1996 the inescapable song was "Macarena," and the ubiquitous adornment was the Cat-in-the-Hat hat—a tall, striped stovepipe accoutrement made famous by Dr. Seuss' feline character. From Massachusetts to California, fairgoers, both old and young, unselfconsciously tromped through animal barns, exhibition halls, and midways with the same striped cartoon hat. Even in Texas staunch cowboys and cowgirls traded their Stetsons for cat wear.

Beneath the tent at the entrance of the Harris County Fair in Houston, Texas, that served as the commercial building, Joseph Dominguez stood behind a table stacked high with the popular headgear. A senior at the University of Houston, Dominguez was a math major with a strong interest in business. A friendly young man with an extraordinary talent for charming people into buying something they would probably wear once, Dominguez told me that he had sold his hats at three fairs—all in Texas—that summer.

"I saw the hats at the Mardi Gras in Galveston," he said. "They were selling like crazy. I had taken a marketing class at school and I realized that a fair would be a great place to sell these things. It's a fad; once everyone has one, that'll be the end of it. By next summer no one will want a Cat-in-the-Hat hat; they'll want something else."

Prize-winning Humboldt County cattle exhibited in the dairy building bring top dollar at auction. Commerical dairies enhance their herds with animals bred from such outstanding show stock. Held in Ferndale, California, for a century, the fair plays a significant role in the local agricultural economy.

country. She can meet all kinds of people and from them learn what it is to be an American."

What is it to be an American? I, like Vershinin's daughter, was born one. Yet it was not until the age of 32, when I embarked upon an itinerary that would take me to county fairs from one end of the nation to the other, that I earnestly began to ponder the question. From Iowa to Alabama, from California to Texas, from Massachusetts to Washington, I gazed upon exhibits, both formal and impromptu, erected to reflect our nation's values. And what are our values? Have they changed since the first Berkshire County Fair? I would have to conclude that they have not. We as a nation are no longer an agrarian society—and, appropriately enough, county fairs are no longer true agricultural events. Why, then, have fairs survived? Why do Americans continue to attend them in ever increasing numbers?

The answer I received from fairgoers across the country was unanimous—the people. People like Jon Zoffka, Tom Morisette, Adam Almeida, Joseph Dominguez, and Dimitri Vershinin, who, though at different points in their lives and from different backgrounds, had come to the fair in pursuit of their own unique American dream.

Fairs are a faithful reflection of not only our small, very human hopes and dreams, but of our constantly evolving society. While all fairs have changed with the times, the most evolved fair that I visited was the Los Angeles County Fair. At that fair's High Tech Expo, hundreds of computer monitors flickered, luring the people of Los Angeles County into cyberspace. The pavilion was a crush of activity as computer wonks shouted for attention and fairgoers elbowed their way onto the information highway. Here, Gordon Ross stood before a bank of computer terminals as he sold the future. Ross, who works for an investment corporation in Vancouver that "mostly deals

with high-end security issues," had developed an Internet software called Net Nanny. It blocks hate literature, pornography, and bomb and drug formulas from reaching your computer terminal. Recognizing key words, phrases, or content that the computer owner does not want accessed from his terminal, the software issues a signal that may terminate the offending application at the users' software control.

A tall man whose intelligent eyes reflected the quick blue flicker of a computer screen, Ross explained, "You can't legislate the Net; it's impossible. So what Net Nanny does is give Mom and Dad a sense of security by protecting the kids from cyberstrangers."

Ross, it would seem, has developed an appropriate product for an America that long ago lost its innocence. Yet, in an era when cyberstrangers lurk in a new, dimensionless world, the county fair, that unassuming American tradition, thrives.

Although agriculture continues to provide a leitmotif, county fairs today are less about farming than they are about community—however it may be defined. It is where we Americans can go, in our ever-increasing diversity, and feel like we belong.

★ ★ ★

Directory of Fairs

UNITED STATES

ALABAMA
Andalusia—Covington County Fair
Anniston—Calhoun County Fair
Birmingham—Alabama State Fair
Boaz—Marshall County Fair
Columbiana—Shelby County Fair
Cullman—Cullman County Fair
Decatur—Tennessee Valley Exposition
  (Morgan County Fair)
Dothan—National Peanut Festival and Fair
Fort Payne—DeKalb County V.F.W.
  Agricultural Fair
Huntsville—Northeast Alabama State Fair
Mobile—Greater Gulf State Fair
Montgomery—Alabama National Fair &
  Agricultural Exposition
Muscle Shoals—North Alabama State Fair
Tuscaloosa—West Alabama State Fair

ALASKA
Delta Junction—Deltana Fair
Fairbanks—Tanana Valley State Fair
Haines—Southeast Alaska State Fair
Ninilchik—Kenai Peninsula State Fair
Palmer—Alaska State Fair

ARIZONA
Casa Grande—Pinal County Fair
Cottonwood—Verde Valley Fair Association
Douglas—Cochise County Fair Association
Flagstaff—Coconino County Fair
Kingman—Mohave County Fair
Parker—La Paz County Fair Association
Phoenix—Arizona State Fair
Phoenix—Maricopa County Fair
Prescott—Yavapai County Fair
Tucson—Pima County Fair
Window Rock—Navajo Nation Fair
Yuma—Yuma County Fair

ARKANSAS
Benton—Saline County Fair
Berryville—Carroll County Fair and
  Livestock Show, Inc.
Blytheville—Mississippi County Fair Association

Camden—Ouachita County Livestock &
  Fair Association, Inc.
Fort Smith—Arkansas-Oklahoma State Fair
Fouke—Miller County Fair
Harrison—Northwest Arkansas District Fair
Jonesboro—Craighead County Fair
Little Rock—Arkansas State Fair &
  Livestock Show
Magnolia—Columbia County Fair and
  Livestock Association
Marvell—Tri County Fair
Searcy—White County Fair
Star City—Lincoln County Fair
Texarkana—Four States Fair

CALIFORNIA
Angels Camp—Calaveras County Fair
Antioch—Contra Costa County Fair
Auburn—Gold Country Fair
Bakersfield—Kern County Fair
Bishop—Eastern Sierra Tri County Fair
Blythe—Colorado River County Fair
Burbank—San Fernando Vally Fair
Calistoga—Napa County Fair
Chico—Silver Dollar Fair
Cloverdale—Cloverdale Citrus Fair
Costa Mesa—Orange County Fair &
  Exposition Center
Crescent City—Del Norte County Fair
Del Mar—Del Mar Fair
Fresno—Fresno Fair
Grass Valley—Nevada County Fair
Gridley—Butte County Fair and Sportsman's
  Expo
Imperial—California Mid-Winter Fair
Indio—Riverside County Fair and National
  Date Festival
King City—Salinas Valley Fair
Lake Perris—Farmers Fair and Expo
Lancaster—Antelope Valley Fair
Lodi—Lodi Grape Festival & Harvest Fair
Madera—Madera District Fair
Napa—Napa Town & Country Fair
Paso Robles—California Mid-State Fair
Petaluma—Sonoma-Marin Fair
Placerville—El Dorado County Fair
Pleasanton—Alameda County Fair

Pomona—Los Angeles County Fair
Porterville—Tulare County Junior Livestock
  Show and Community Fair
Red Bluff—Tehama District Fair
Ridgecrest—Desert Empire Fair
Roseville—Placer County Fair
Sacramento—California State Fair
San Bernardino—National Orange Show
San Francisco—Grand National Rodeo, Horse,
  and Stock Show
San Francisco—San Francisco Fair
San Jose—Santa Clara County Fair
San Mateo—San Mateo County Fair &
  Floral Festival
Santa Barbara—Santa Barbara Fair and Expo
Santa Maria—Santa Barbara County Fair
Santa Rosa, Sonoma County Fair & Exposition
Sonora—The Mother Lode Fair
Stockton—San Joaquin County Fair
Tulare—Tulare County Fair
Vallejo—Solano County Fair
Ventura—Ventura County Fair
Victorville—San Bernardino County Fair
Yreka—Siskiyou Golden Fair
Yuba City—Yuba Sutter Fair

COLORADO
Black Hawk—Gilpin County Fair
Calhan—El Paso County Fair
Denver—Arapahoe County Fair
Denver—National Western Stock Show
  & Rodeo
Durango—LaPlata County Fair
Eagle—Eagle County Fair & Rodeo
Estes Park—Rooftop Fair & Rodeo
Grand Junction—Mesa County Fair
Greeley—Weld County Fair
Henderson—Adams County Fair & Rodeo
Longmont—Boulder County Fair &
  Livestock Show
Loveland—Larimer County Fair & Rodeo
Poncha Springs—New Old Fashioned Chaffee
  County Fair
Pueblo—Colorado State Fair
Rocky Ford—Arkansas Valley Fair and
  Exposition
Sterling—Logan County Fair and Rodeo

Source: Reprinted with permission of International Association of Fairs and Expositions 1997.

CONNECTICUT
Berlin—Berlin Fair
Bethlehem—Bethlehem Fair
Bridgewater—Bridgewater Country Fair
Brooklyn—Brooklyn Fair
Durham—Durham Agricultural Fair
Goshen—Connecticut Agricultural Fair, Inc.
Goshen—Goshen Agricultural Society
Harwinton—Harwinton Fair
Hebron—Hebron Harvest Fair
Lebanon—Lebanon Country Fair
North Haven—North Haven Fair
North Stonington—North Stonington
    Agricultural Fair, Inc.
Riverton—Riverton Fair
Somers—Union Agricultural Society, Inc.
    (Four Town Fair)
South Windsor—Wapping Fair
Terryville—Terryville Country Fair, Inc.
Wolcott—Wolcott Lions Country Fair
Woodstock—Woodstock Fair

DELAWARE
Harrington—Delaware State Fair

FLORIDA
Auburndale—Florida Citrus Festival and
    Polk County Fair
Bonifay—Holmes County Fair
Bushnell—Sumter County Fair
Callahan—Northeast Florida Fair
Clewiston—Hendry County Fair
Dade City—Pasco County Fair
De Funiak Springs—Walton County Fair
De Land—Volusia County Fair &
    Youth Show
Eustis—Lake County Fair
Fort Myers—Southwest Florida and Lee
    County Fair Association
Fort Pierce—St. Lucie County Fair
Fort Walton Beach—Greater Okaloosa
    County Fair
Gainesville—Alachua County Fair
Green Cove Springs—Clay County Fair
Homestead—Country Fair of South Dade
Inverness—Citrus County Fair
Jacksonville—Greater Jacksonville Agricultural
    Fair
Jasper—Hamilton County Fair
Key West—Monroe County Fair
Kissimmee—Kissimmee Valley Livestock
    Show and Fair
Lake City—Columbia County Fair
Live Oak—Suwannee County Fair & Youth
    Livestock Show

Mac Clenny—Baker County Fair
Marianna—Jackson County Agricultural
    Exposition
Miami—Dade County Fair & Exposition
Milton—Santa Rosa County Fair
Naples—Collier County Fair
Orlando—Central Florida Fair
Palmetto—Manatee County Fair
Panama City—Bay County Fair
Pensacola—Pensacola Interstate Fair
Pinellas Park—Pinellas County Fair and
    Florida Music Festival
Plant City—Florida Strawberry Festival
    and Fair
Port Charlotte—Charlotte County Fair
St. Augustine—St. John's County Fair
Sarasota—Sarasota County Fair
Sebring—Highlands County Fair
Starke—Bradford County Fair
Stuart—Martin County Fair
Tallahassee—North Florida Fair
Tampa—Florida State Fair
Vero Beach—Indian River County Firefighters
    Fair
Wauchula—Hardee County Fair
West Palm Beach—South Florida Fair &
    Palm Beach County Expositions, Inc.

GEORGIA
Albany—Exchange Club Fair of Southwest
    Georgia
Augusta—Exchange Club of Augusta Fair
Bainbridge—Decatur County Fall Festival
Brunswick—Brunswick Exchange Club
    Agricultural Fair
Columbus—Chattahoochee Valley Fair
Cumming—Cumming County Fair and Festival
Dalton—North Georgia Fair
Gray—Jones County Lions Club Fair
Hiawassee—Georgia Mountain Fair
Jackson—Butts County Fair
Lawrenceville—Gwinnett County Livestock and
    Fair Association
Macon—Georgia State Fair
Marietta—North Georgia State Fair
Milledgeville—Oconee Area Fair
Perry—Georgia National Fairgrounds &
    Agricenter
Rome—Coosa Valley Fair
Savannah—Savannah Exchange Club Fair
    Association, Inc.
Statesboro—Kiwanis Ogeechee Fair
Tifton—Coastal Plains Agricultural Fair
Waycross—Okefenokee Agricultural Fair
Winder—Winder Lions Club Fair

HAWAII
Honolulu—Hawaii State Farm Fair
Kahului—Maui County Fair

IDAHO
Blackfoot—Eastern Idaho State Fair
Boise—Western Idaho Fair
Caldwell—Canyon County Fair
Coeur D'Alene—North Idaho Fair
Filer—Twin Falls County Fair
Gooding—Gooding County Fair & Rodeo
Grace—Caribou County Fair and Rodeo
Jerome—Jerome County Fair & Rodeo
Lewiston—Nez Perce County Fair
Pocatello—Bannock County Fair
Rexburg—Madison County Fair
Sandpoint—Bonner County Fair
Shoshone—Lincoln County Fair and Rodeo

ILLINOIS
Albion—Edwards County Fair
Aledo—Mercer County Fair
Altamont—Effingham County Fair
Anna—Southern Illinois Fair
Belleville—St. Clair County Fair
Belvidere—Boone County Fair
Benton—Frankline County 4-H Fair
Bloomington—McLean County Fair
Carlinville—Macoupin County Fair &
    Agri. Assn., Inc.
Carmi—White County Fair
Carrollton—Greene County Agricultural Fair
Cerro Gordo—Piatt County Junior Fair
Crescent City—Iroquois County Agricultural
    and 4-H Club Fair
Decatur—Decatur-Macon County Fair
Du Quoin—Du Quoin State Fair
East Moline—Rock Island County Fair
Elizabeth—Elizabeth Community Fair
Fairbury—The Fairbury Fair, Inc.
Farmer City—Farmer City Association
Fisher—Fisher Coummunity Fair
Georgetown—Georgetown Fair Association
Grayslake—Lake County Fair
Greenville—Bond County Fair
Griggsville—Western Illinois Fair
Henry—Marshall-Putnam County Fair
Highland—Madison County Fair
Jacksonville—Morgan County Agricultural
    Fair Association
Jerseyville—Jersey County Fair
Kankakee—Kankakee County Fair &
    Expo, Inc.
Knoxville—Knox County Fair
Lincoln—Logan County Fair

Macomb—McDonough County 4-H and
   Junior Fair
Marshall—Clark County Fair
Martinsville—Martinsville Agricultural Fair
Melvin—Ford County Fair
Mendon—Adams County Fair
Mendota—Tri County Fair
Milledgeville—Caroll County Fair
Morris—Grundy County Fair
Morrison—Whiteside County Fair
Nashville—Nashville, Washington County Fair
New Berlin—Sangamon County Fair
Oblong—Crawford County Fair
Olney—Richland County Fair
Oregon—Ogle County Fair
Paris—Edgar County Fair
Pecatonica—Winnebago County Fair
Peoria—Heart of Illinois Fair
Peotone—Will County Fair Association
Petersburg—Menard County Fair
Pleasant Hill—Pike County Fair
Pontiac—Livingston County Agricultural Fair
Princeton—Bureau County Fair
Pulaski—Pulaski Civic & Community
   Fair Association, Inc.
Ridgway—Gallatin County Fair
Roseville—Warren County Fair
Rushville—Schuyler County Fair
St. Charles—Kane County Fair &
   Industrial Exposition
Salem—Marion County Agricultural Fair
Sandwich—DeKalb County Fair
Springfield—Illinois State Fair
Taylorville—Christian County Agricultural
   Fair Association
Urbana—Champaign County Fair
Virginia—Cass County Fair
Warren—Jo Daviess County Agricultural Society
Waterloo—Monroe County Fair
Wheaton—DuPage County Fair
Woodstock—McHenry County Fair Association

INDIANA
Anderson—Anderson Free Fair
Brownstown—Jackson County Fair
Corydon—Harrison County Fair
Crown Point—Lake County Fair
Elnora—Daviess County Fair
English—Crawford County 4-H Fair &
   English Reunion
Franklin—Johnson County 4-H and
   Agricultural Fair
Goshen—Elkhart County 4-H Fair
Greentown—Howard County 4-H Fair
Indianapolis—Indiana State Fair

Indianapolis—Marion County Agricultural
   and 4-H Club Fair
Lagrange—Lagrange County 4-H Fair
La Porte—La Porte County Fair
Marion—Grant County 4-H Fair Association
Muncie—Lions Delaware County Fair
Portland—Jay County Fair
Rushville—Rush County Agricultural
   Association
Salem—Washington County Fair
Shelbyville—Shelby County Fair
South Bend—St. Joseph County 4-H Fair
Spencer—Owen County Fair
Valparaiso—Porter County Fair
Warsaw—Kosciusko County 4-H and
   Community Fair

IOWA
Adel—Dallas County Fair
Alta—Buena Vista County Agricultural Society
Audubon—Audubon County Fair
Boone—Boone County Fair
Cedar Rapids—All Iowa Fair
Cherokee—Cherokee County Fair
Colfax—Jasper County 4-H and EFA Fair
Columbus Junction—Louisa County Fair
   Association
Council Bluffs—Westfair
Cresco—The "Mighty" Howard County Fair
Davenport—Mississippi Valley Fair
Decorah—Winneshiek County Fair
Denison—Crawford County Fair
Des Moines—Iowa State Fair
Des Moines—Polk County 4-H and EFA Fair
Dubuque— Dubuque County Fair
Eldon—Wapello County Regional Fair
Harlan—Shelby County Fair
Ida Grove—Ida County Fair
Independence—Buchanan County Fair
Indianola—Warren County Agricultural Fair
Leon—Decatur County Fair
Manchester—Delaware County Fair
Mason City— North Iowa Fair
Missouri Valley—Harrison County Fair
Monticello—Great Jones County Fair
Mt. Pleasant—Midwest Old Threshers Reunion
Moville—Woodbury County Fair
Oskaloosa—Southern Iowa Fair
Red Oak—Montgomery County Agricultural
   Society
Spencer—Clay County Fair
Vinton—Benton County Fair
Waterloo—National Cattle Congress
Waverly—Bremer County Fair Association
West Burlington—Des Moines County Fair

West Liberty—Muscatine County Fair
West Union—Fayette County Fair
Winterset—Madison County Livestock and
   Fair Association

KANSAS
Abilene—Central Kansas Free Fair
Belleville—North Central Kansas Free Fair
Blue Rapids—Marshall County Stock Show
   and Fair
Coffeyville—Inter-State Fair
Colby—Thomas County Free Fair
Concordia—Cloud County Fair Association
Garden City—Finney County Fair
Garnett—Anderson County Fair
Hays—Ellis County Fair
Hillsboro—Marion County Fair
Hugoton—Steven County Fair Association
Hutchinson—Kansas State Fair
Kansas City—Wyandotte County Fair
La Crosse—Rush County Fair
Liberal—Five State Free Fair
Ness City—Ness County Fair
Olathe—Johnson County Old Settlers
Oswego—Labette County Fair
Pratt—Pratt County Fair Association
Scott City—Scott County Free Fair
Sedan—Chautauqua County Fair
Stockton—Rooks County Free Fair
Syracuse—Hamilton County Fair
Ulysses—Grant County Free Fair
Winfield—Cowley County Fair

KENTUCKY
Ashland—Boyd County Fair, Inc.
Crittenden—Grant County Fair and Horse Show
Ewing—Ewing-Fleming County Fair
Germantown—Germantown Fair
Grayson—Carter County Fair
Hartford—Ohio County Fair Association
Lexington—Lexington Lions Bluegrass Fair
London—Laurel County Fair
Louisville—Kentucky State Fair
Louisville—North American International
   Livestock Exposition
Murray—Murray-Calloway County Fair
Paducah—McCracken County Fair

LOUISIANA
Abbeville—Lousiana Cattle Festival
Alexandria—Rapides Parish Fair
Baton Rouge—Greater Baton Rouge State Fair
Bernice—Corney Creek Porkfest
Breaux Bridge—Breaux Bridge Crawfish Festival
De Ridder—Beauregard Parish Fair

Franklinton—Washington Parish Watermelon
   Festival Association
Lacombe—Bayou Lacombe Crab Festival
Lafayette—Cajun Heartland State Fair
Lafayette—SummerFest
Livingston—Livingston Parish Fair Association
Many—Sabine Parish Fair
Shreveport—State Fair of Louisiana
Sulphur—Calcasieu Cameron Fair
Thibodaux—Thibodaux Firemen's Fair
West Monroe—Ark-La-Mis Fair

MAINE
Bangor—Bangor State Fair
Dover-Foxcroft— Piscataquis Valley Fair
   Association
Farmington—Franklin County Agricultural
   Society
Fryeburg—Fryeburg Fair
Monmouth—Cochnewagen Agricultural
   Association
Skowhegan—Skowhegan State Fair
South Hiram—Ossipee Valley Fair
Topsham—Topsham Fair
West Cumberland—Cumberland County Fair

MARYLAND
Bel Air—Harford County Farm Fair
Centreville—Queen Anne's County Fair
Crownsville—Anne Arundel County Fair
Cumberland—Allegany County Fair
Damascus—Damascus Community Fair
Fair Hill—Cecil County Fair
Frederick—Great Frederick Fair
Gaithersburg—Montgomery County
   Agricultural Fair
Jefferson—Maryland Sheep and Wool Festival
La Plata—Charles County Fair
Leonardtown—St. Mary's County Fair
Mc Henry—Garrett County Fair
Middletown—Middletown-Braddock
   Community Show
Pocomoke City—Great Pocomoke Fair
Prince Frederick—Calvert County Fair, Inc.
Salisbury—Wicomico County Fair
Timonium—Baltimore County 4-H Fair
Timonium—Eastern National Livestock Show
Timonium—Maryland State Fair
Upper Marlboro—Prince George's County Fair
Westminster—Carroll County 4-H FFA Fair

MASSACHUSETTS
Belchertown—Belchertown Fair
Brockton—Brockton-Middleboro
   Agricultural Fair

Cummington—Cummington Fair
Falmouth—Barnstable County Fair
Greenfield—Franklin County Fair
Marshfield—Marshfield Fair
Northampton—Three County Fair
Spencer—Spencer Fair
Taunton—Rehoboth Fair, Inc.
Topsfield—Topsfield Fair
West Springfield—Eastern States Exposition
West Tisbury—Martha's Vineyard
   Agricultural Society
Westfield—Westfield Fair

MICHIGAN
Adrian—Lenawee County Fair
Allegan—Allegan County Fair
Ann Arbor—Washtenaw County 4-H
   Youth Fair
Belleville—Wayne County Fair
Berrien Springs—Berrien County Youth Fair
Caro—Tuscola County Fair
Cassopolis—Cass County Fair Association
Centreville—St. Joseph County Grange Fair
Coldwater—Branch County 4-H Fair
Corunna—Shiawassee County Fair
Davisburg—Oakland County 4-H Fair
Detroit— Michigan State Fair and
   Exposition Center
Escanaba—Upper Peninsula State Fair
Fowlerville—Fowlerville Fair
Hartford—Van Buren County Youth Fair
Hastings—Barry County Fair
Hillsdale—Hillsdale County Agricultural Fair
Holland—Ottawa County Fair
Hudsonville—Hudsonville Community Fair
Ionia—Ionia Free Fair
Jackson—Jackson County Fair
Kalamazoo—Kalamazoo County
   Agricultural Fair
Lake Odessa—Lake Odessa Fair
Manchester—Manchester Community Fair
Mason—Ingham County Fair
Midland—Midland County Agricultural &
   Horticultural Fair
Monroe—Monroe County Fair
Moran—Mackinac County Fair
Mt. Morris—Genesee County Fair
Pelkie—Baraga County Fair
Petoskey—Emmet County Fair
Saginaw—Saginaw Fair
Stephenson—Menominee County Fair
   Association
Traverse City—Northeastern Michigan Fair
   Association
West Branch—Ogemaw County Fair

MINNESOTA
Ada—Norman County Fair
Albert Lea—Freeborn County Fair
Anoka—Anoka County Suburban Fair
Arlington—Sibley County Fair
Austin—Mower County Fair
Barnesville—Clay County Fair and
   Agricultural Association
Barnum—Carlton County Fair
Bird Island—Renville County Fair
Caledonia—Houston County Fair
Cambridge—Isanti County Fair
Cannon Falls—Cannon Valley Fair
Corcoran—Hennepin County Fair
Crookston—Red River Valley Winter Shows
Duluth—South St. Louis County Fair
Elk River—Sherburne County Fair
Faribault—Rice County Agricultural Fair
Farmington—Dakota County Fair
Fergus Falls—West Ottertail County Fair
Goodhue—Goodhue County Fair
Grand Rapids—Itasca County Fair
Hallock—Kittson County Fair
Hibbing—St. Louis County Fair
Howard Lake—Wright County Fair
Jackson—Jackson County Fair
Kasson—Dodge County Fair
Lake Elmo—Washington County Fair
Little Falls—Morrison County Fair
Littlefork—Northern Minnesota
   District Fair
Luverne—Rock County Fair
Madison—Lac Qui Parle County Fair
New Ulm—Brown County Fair
Owatonna—Steele County Free Fair
Perham—East Ottertail County Fair
Pillager—Cass County Agricultural Society
Pine City—Pine County Fair
Preston—Fillmore County Fair
Princeton—Mille Lacs County Fair
Rochester—Olmsted County Fair
Roseau—Roseau County Fair
Rush City—Chisago County Agricultural
   Society
St. Charles—Winona County Fair
St. James—Watonwan County Fair
St. Paul—Minnesota State Fair
St. Paul—Ramsey County Fair
Slayton—Murray County Fair
Two Harbors—Lake County
   Agricultural Fair
Waconia—Carver County Fair
Wadena—Wadena County Fair
Waseca—Waseca County Agricultural Society
Willmar—Kandiyohi County Fair

MISSISSIPPI

Biloxi—Mississippi Coast Fair & Exposition
Jackson—Mississippi State Fair
Laurel—South Mississippi Fair
Meridian—Mississippi-Alabama State Fair
Philadelphia—Neshoba County Fair
Verona—North Mississippi Fair
Yazoo City—Yazoo County Fair

MISSOURI

Cape Girardeau—Southeast Missouri
  District Fair
Chesterfield—St. Louis County Fair
Columbia—Boone County Fair and
  Horse Show
Galena—Stone County Fair
Kansas City—American Royal Livestock,
  Horse Show & Rodeo
Lamar—Lamar Free Fair
Laurie—Hillbilly Fair
Lebanon—Laclede County Community Fair
Marshfield—Webster County Fair
Montgomery City—Montgomery County Fair
Owensville—Gasconade County Fair
Palmyra—Marion County Fair
Pleasant Hill—Cass County Fair
Potosi—Washington County Fair
Richmond—Ray County Fair
Rolla—Central Missouri Regional Fair, Inc.
Sedalia—Missouri State Fair
Springfield—Ozark Empire Fair
Tracy—Platte County Fair
Troy—Lincoln County Fair
Vandalia—Vandalia Area Fair
Washington—Washington Town and
  Country Fair
Wellington—Wellington Community
  Fair, Inc.
West Plains—Heart of the Ozarks Fair

MONTANA

Baker—Fallon County Fair
Billings—MetraPark
Billings—Northern International Livestock
  Exposition and Rodeo
Bozeman—Gallatin County Fair
Bozeman—Montana Winter Fair, Inc.
Butte—Butte Silver Bow County Fair
Chinook—Blaine County Fair
Forsyth—Rosebud-Treasure County Fair
Fort Benton—Chouteau County Fair
Glasgow—Northeast Montana Fair
Glendive—Dawson County Fair
Great Falls—Montana State Fair
Hamilton—Ravalli County Fair

Havre—Great Northern Fair
Helena—Last Chance Fair and Stampede
Kalispell—Northwest Montana Fair & Rodeo
Lewistown—Central Montana Fair
Livingston—Park County Fair
Miles City—Eastern Montana Fair
Missoula—Western Montana Fair
Plains—Sanders County Fair
Plentywood—Sheridan County Fair
Sidney—Richland County Fair & Rodeo
Twin Bridges—Madison County Fair

NEBRASKA

Aurora—Hamilton County Fair
Beatrice—Gage County Fair & Exposition
Beaver City—Furnas County Fair
Bloomfield—Knox County Fair
Central City—Merrick County Agricultural Fair
Chambers—Holt County Fair
Columbus—Platte County Fair
Eustis—Eustis Agrcultural Society
Fullerton—Nance County Fair
Geneva—Fillmore County Fair
Hastings—Adams County Agricultural Society
Hayes Center—Hayes County Fair Association
Hemingford—Box Butte County Fair
Humboldt—Richardson County Free Fair
Imperial—Chase County Fair and Exposition
Johnstown—Brown County Fair and Rodeo
Kearney—Buffalo County Fair
Kearney— Gateway Farm Expo
Kearney—Nebraska Cattlemen's Classic
Lexington—Dawson County Fair
Lincoln—Lancaster County Fair
Lincoln—Nebraska State Fair
Madison—Madison County Fair & Rodeo
Mc Cook—Red Willow County Fair
Minden—Kearney County Fair
Mitchell—Scotts Bluff County Fair
Neligh—Antelope County Fair
North Platte—Lincoln County Fair
Oakland—Burt County Agricultural Society
Ogallala—Keith County Fair
Omaha—Ak-Sar-Ben 4-H Livestock Exposition
Omaha—Douglas County Fair
St. Paul—Howard County Agricultural Society
South Sioux City—Greater Siouxland Fair &
  Rodeo/Atokad Park
Spalding—Greeley County Free Fair
Stockville—Frontier County Fair
Valentine—Cherry County Fair and Rodeo
Wayne—Wayne County Fair
Weeping Water—Cass County Agricultural
  Society
York—York County Fair

NEVADA

Elko—Elko County Fair & Livestock Show
Las Vegas—Las Vegas Jaycees State Fair
Logandale—Clark County Fair
Mills Park—RSVP Spring Fun Fair
Pahrump—Pahrump Harvest Festival & Fair
Panaca—Lincoln County Fair
Reno—Nevada State Fair
Winnemucca—Tri County Fair Board
Yerington—Lyon County Fair

NEW HAMPSHIRE

Contoocook—Hopkinton State Fair
Cornish—Cornish Fair
Deerfield—Deerfield Fair
Keene—Cheshire Fair Association
Lancaster—Coos-Essex Agricultural Fair
North Haverhill—North Haverhill Fair
  Association
Rochester—Rochester Fair
Stratham—Stratham Fair

NEW JERSEY

Augusta—Sussex County Farm & Horse Show
Belvidere—Warren County Farmers' Fair
Berkeley Township—Ocean County Fair
Cherry Hill—New Jersey State Fair
Chester—Morris County 4-H Fair
Flemington—Flemington Agricultural Fair
Lumberton—Burlington County Farm Fair, Inc.
Millville—Cumberland County Co-Op Fair
Mullica Hill—Gloucester County 4-H Fair

NEW MEXICO

Albuquerque—New Mexico State Fair
Farmington—San Juan County Fair
Las Cruces—Southern New Mexico State Fair
Lovington—Lea County Fair and Rodeo
Roswell—Eastern New Mexico State Fair

NEW YORK

Afton—Afton Fair
Altamont—Altamont Fair
Angelica—Allegany County Fair
Ballston Spa—Saratoga County Fair
Batavia—Genesee County Agricultural Fair
Bath—Steuben County Fair
Belmont Park—Greater New York Fair
  and Festival
Boonville—Boonville Fair
Brookfield—Madison County Fair
Caledonia—Caledonia Fair
Canandaigua—Ontario County Agricultural
  Society, Inc.
Chatham—Columbia County Fair

Cobleskill—Cobleskill "Sunshine" Fair

Dunkirk—Chautauqua County Agricultural Fair

Elmira-Horseheads—Chemung County Fair

Fonda—Montgomery County Agricultural
Society

Frankfort—Herkimer County Fair

Gouverneur—Gouverneur & St. Lawrence
County Fair

Greenwich—Washington County Fair, Inc.

Hamburg—Erie County Fair and Exposition

Hemlock—Hemlock Fair

Little Valley—Cattaraugus County Agricultural
Society

Lockport—Niagara County Fair

Lowville—Lewis County Agricultural Society

Malone—Franklin County Agricultural Society

Middletown—Orange County Fair

Morris—Otsego County Fair Association

New Paltz—Ulster County Fair

Norwich—Chenango County Fair

Penn Yan—Yates County Fair

Plattsburgh—The Agricultural & Industrial Fair
of Clinton County, Inc.

Rhinebeck—Dutchess County Fair

Rochester—Monroe County Fair

Sandy Creek—Oswego County Fair

Schaghticoke—Rensselaer County Agricultural
& Horticultural Society

Syracuse—New York State Fair

Walton—Delaware Valley Agricultural Society

Watertown—Jefferson County Fair

Weedsport—Cayuga County Fair

Westport—Essex County Agricultural Society

Whitney Point—Broome County
Agricultural Fair

Yonkers—Greater Westchester County Fair
& Exposition

Yorktown Heights—Yorktown Grange Fair
Association

NORTH CAROLINA

Albemarle—Stanly County Fair Association

Burlington—Alamance County Agricultural Fair

Concord—Cabarrus County Agricultural Fair

Eden—Eden Fair

Edenton—Chowan County Fair

Fayetteville—Cumberland County Fair

Fletcher—North Carolina Mountain State Fair

Fort Bragg—Fort Bragg Fair

Goldsboro—Wayne County Agricultural Fair

Greenville—Pitt County American Legion
Agricultural Fair

Henderson—Vance County Regional Fair

Hickory—Hickory American Legion Fair

Jacksonville—Onslow County Fair

Kings Mountain—Bethware Community Fair

Kinston—Lenoir County Fair

Lenoir—Caldwell County Agricultural Fair

Lexington—Davidson County Agricultural Fair

Lumberton—Robeson County Fair

Mt. Airy—Surry County Agricultural Fair

New Bern—Craven, Pimlico, Carteret County
Agricultural Fair

Newland—Avery County Agricultural and
Horticultural Fair

Raleigh—North Carolina State Fair

Roanoke Rapids—Halifax-Northampton
Agri Fair

Rocky Mount—Rocky Mount Fair

Salisbury—Rowan County Agricultural &
Industrial Fair

Shelby—Cleveland County Fair

Wilmington—New Hanover County Fair
Association

Winston-Salem—Dixie Classic Fair

Yanceyville—Caswell County Agricultural Fair

NORTH DAKOTA

Amidon—Slope Farmers Fair

Beach—Golden Valley County Fair

Beulah—Mercer County Fair

Bismarck—Expo 97

Bismarck—Missouri Valley Fair

Bottineau—Bottineau County Fair

Bowman—Bowman County Fair Association

Carrington—Foster County Fair

Cooperstown—Griggs County Fair

Crosby—Divide County Fair

Devils Lake—Ramsey County Fair

Dickinson—Roughrider Days

Ellendale—Dickey County Fair

Fessenden—Wells County Fair

Flaxton—Burke County Fair Association

Forman—Sargent County Fair Association

Grand Forks—Greater Grand Forks Fair and
Exhibition Association, Inc.

Hamilton—Pembina County Fair Association

Hettinger—Adams County Achievement Days

Jamestown—Stutsman County Fair

Lisbon—Ransom County Fair

Minot—North Dakota State Fair

Mott—Hettinger County Fair Association

New Salem—Morton County Fair

Rugby—Pierce County Fair

Underwood—McLean County Fair

Valley City—North Dakota Winter Show

Watford City—McKenzie County Fair

West Fargo—Red River Valley Fair

Williston—Upper Missouri Valley Fair

Wishek—Tri-County Fair

OHIO

Berea—Cuyahoga County Fair

Bucyrus—Crawford County Agricultural Society

Canfield—Canfield Fair

Chillicothe—Ross County Fair

Cincinnati—Hamilton County Carthage Fair

Columbus—Ohio State Fair

Coshocton—Coshocton County Fair

Dayton—Montgomery County Fair

Delaware—Delaware County Fair

Eaton—Preble County Fair

Fremont—Sandusky County Fair

Greenville—The Great Darke County Fair

Hamilton—Butler County Fair

Hillsboro—Highland County Agricultural
Society

Kenton—Hardin County Fair

Lancaster—Fairfield County Fair

Lima—Allen County Fair

Lucasville—Scioto County Agricultural Fair

Marietta—Washington County Fair

Marion—Marion County Fair

Marysville—Union County Agricultural Society

Maumee—Lucas County Agricultural Fair

Mc Arthur—Vinton County Junior Fair

Mc Connelsville—Morgan County Fair

Medina—Medina County Fair

New Lexington—Perry County Fair

Oak Harbor—Ottawa County Fair

Owensville—Clermont County Fair

Painesville—Lake County Fair

Randolph—Portage County Randolph Fair

Springfield—Clark County Fair

Tallmadge—Summit County Fair

Van Wert—Van Wert County Agricultural
Society

Wapakoneta—Auglaize County Fair

Washington C.H.—Fayette County Fair

Wauseon—Fulton County Fair

Wellington—Lorain County Fair

West Union—Adams County Fair

Wooster—Wayne County Fair

Xenia—Greene County Fair

Zanesville—Muskingum County Fair

OKLAHOMA

El Reno—Canadian County Free Fair

Enid—Garfield County Fair

Enid—Northwest District Junior Livestock
Show

Guymon—Panhandle Exposition

Mc Alester—Pittsburg County Free Fair

Norman—Cleveland County Free Fair

Oklahoma City—State Fair of Oklahoma

Pryor—Mayes County Free Fair

Stillwater—Payne County Fair
Tulsa—Tulsa State Fair
Watonga—Blaine County Free Fair

OREGON
Astoria—Clatsop County Fair
Canby—Clackamas County Fair
Central Point—Jackson County Fair
Corvallis—Benton County Fair
Eugene—Lane County Fair
Grants Pass—Josephine County Fair
Hillsboro—Washington County Fair
Klamath Falls—Klamath County Fair
Madras—Jefferson County Fair
Mc Minnville—Yamhill County Fair
Portland—Multnomah County Fair
Prineville—Crook County Fair
Redmond—Deschutes County Fair
Rickreall—Polk County Fair
Roseburg—Douglas County Fair
Salem—Oregon State Fair
Tillamook—Tillamook County Fair

PENNSYLVANIA
Albion—Albion Area Fair
Allentown—The Great Allentown Fair
Altoona—Sinking Valley Farm Show
Arendtsville—South Mountain Community
    and Fair Association
Bangor—Blue Valley Farm Show
Bedford—Bedford County Fair
Bloomsburg—Bloomsburg Fair
Brookville—Jefferson County Fair
Butler—Butler Farm Show, Inc.
Centre Hall—Centre County Grange Fair
Clearfield—Clearfield County Fair
Commodore—Cookport Fair
Dallas—Luzerne County Fair
Dayton—Dayton A&M Fair
Denver—Denver Community Fair
Dillsburg—Dillsburg Community Fair
Ebensburg—Cambria County Fair
Elizabethtown—Elizabethtown Fair
Emporioum—Cameron County Fair
Farmington—Mountain Area Community Fair
Forksville—Sullivan County Agricultural Society
Franklin—Venango County 4-H Fair
Gilbert—West End Fair
Gratz—Gratz Fair
Greensburg—Westmoreland Agricultural Fair
Harford—Harford Fair
Harrisburg—Keystone International Livestock
    Exposition
Harrisburg—Pennsylvania Farm Show
Home—Ox Hill Community Agricultural Fair

Honesdale—Wayne County Fair
Hookstown—Hookstown Fair
Hughesville—Lycoming County Fair
Huntingdon—The Huntingdon County Fair
Indiana—Indiana County Fair
Jamestown—Jamestown Community Fair
Kersey—Elk County Fair
Kutztown—Kutztown Fair
Lampeter—West Lampeter Community Fair
Laurelton—Union County West End Fair
Mackeyville— Clinton County Fair
Manheim—Manheim Community Farm
    Show and Fair
Meadville—Crawford County Fair
Mercer—Jefferson Township Fair
Meshoppen Township—Wyoming County Fair
Meyersdale—Somerset County Fair
Nazareth—Plainfield Farmers Fair
New Bethlehem—Clarion County Fair
New Castle—Lawrence County Fair
Newfoundland—Greene Dreher Sterling Fair
Newport—Perry County Fair
Philadelphia—Pennsylvania Fair
Philadelphia—Philadelphia County Fair
Pittsburgh—Warren County Fair
Pittsfield—Warren County Fair
Port Royal—Juniata County Fair
Prospect—Butler Fair and Agricultural
    Assocation
Reading—Reading Fair
Rochester—Big Knob Grange Fair
Schnecksville—Schnecksville Community Fair
Smethport—McKean County Fair
Spartansburg—Spartansburg Community Fair
Stoneboro—The Great Stoneboro Fair
Summit Station—Schuylkill County Fair
Sykesville—Sykesville Ag & Youth Fair
Troy—Troy Fair
Uniontown—Fayette County Fair
Uniontown—Uniontown Poultry Show
Unionville—Unionville Coummunity Fair
Washington—Washington County Agricultural
    Fair, Inc.
Waterford—Waterford Community Fair
Wattsburg—Erie County Fair
Westover—Haromony Grange Fair
Whitneyville/Wellsboro—Tioga County Fair
Wrights Town—Middletown Grange Fair
York—York Inter-State Fair

RHODE ISLAND
Richmond—Washington County Fair

SOUTH CAROLINA
Aiken—Aiken Jaycee County Fair

Anderson—Anderson County Fair
Charleston—Coastal Carolina Fair
Columbia—South Carolina State Fair
Florence—Eastern Carolina Agricultural Fair
Georgetown—Georgetown County Exposition
    and Fair
Greenville—Upper South Carolina State Fair
Lancaster—Lancaster County Fair
Laurens—Laurens County Fair
Moncks Corner— Berkeley County Fair
Orangeburg—Orangeburg County Fair
Rock Hill—York County Fair
Spartanburg—Piedmont Interstate Fair
Sumter—Sumter County Fair
Union—Union County Agricultural Fair
    Association

SOUTH DAKOTA
Aberdeen—Brown County Fair & 4-H Show
Huron—South Dakota State Fair
Onida—Sully County Fair
Rapid City—Central States Fair
Sioux Falls—Sioux Empire Fair
Webster—Day County Fair

TENNESSEE
Clinton—Anderson County Fair
Crossville—Cumberland County Fair
Dickson—Dickson County Fair
Dyersburg—Dyer County Fair
Fayetteville—Lincoln County Fair
Gallatin—Sumner County Fair Association
Gray—Appalachian Fair
Greeneville—Greene County Fair
Huntingdon—Carroll County Fair
Jackson—West Tennessee State Fair
Jamestown—Fentress County Fair
Knoxville—Tennessee Valley Agricultural and
    Industrial Fair
Lawrenceburg—Middle Tennessee District Fair
Lebanon—Wilson County Fair
Lexington—Heneron County Free Fair
Manchester—Coffee County Fair Association
Mc Minnville—Warren County Agricultural &
    Livestock Fair
Memphis—Mid-South Fair
Nashville—Tennessee State Fair
Newport—Cocke County Agricultural and
    Industrial Fair
Oneida—Scott County Fair
Savannah—Hardin County Fair
Sevierville—Sevier County Fair
Sparta—White County Fair
Trenton—Gibson County Fair
Union City—Obion County Fair

Waverly—Humphreys County Ag &
Industry Fair

TEXAS

Abilene—West Texas Fair & Rodeo
Amarillo—Amarillo Tri State Fair
Angleton—Brazoria County Fair
Austin—Austin-Travis County Livestock Show
& Rodeo
Bay City—Bay City Lions Club Rice Festival
Bay City—Matagorda County Fair and
Livestock Show
Beaumont—South Texas State Fair
Bellville—Austin County Fair
Belton—Central Texas State Fair
Big Spring—Howard County Fair
Association, Inc.
Brenham—Washington County Fair
Caldwell—Burleson County Fair
Coldspring—San Jacinto County Fair
Conroe—Montgomery County Fair
Crosby—Crosby Fair and Rodeo
Dallas—State Fair of Texas
Denton—North Texas State Fair
Fort Worth—Southwestern Exposition &
Livestock Show
Fredericksburg—Gillespie County Fair
Greenville—Hunt County Fair Association, Inc.
Hondo—Medina County Fair
Houston—Harris County Fair
Houston—Houston Livestock Show & Rodeo
Huntsville—Walker County Fair Association
Johnson City—Blanco County Fair & Rodeo
Killeen—Central Texas Exposition, Inc.
La Grange—Fayette County Country Fair
Liberty—Trinity Valley Exposition
Longview—Longview Jaycees Gregg County Fair
& Exposition
Longview—Northeast Texas Regional Fair
Lubbock—Panhandle-South Plains Fair
Lufkin—Texas Forest Festival
Mercedes—Rio Grande Valley Livestock Show
Mt. Pleasant—Titus County Fair
Nacogdoches—Piney Woods Fair
Odessa—Permian Basin Fair & Exposition
Paris—Red River Valley Fair & Expo
Poteet—Poteet Strawberry Festival
Rio Grande City—Starr County Fair Association
Rosenberg—Fort Bend County Fair
San Angelo—San Angelo Stock Show and Rodeo
Association
San Antonio—San Antonio Livestock
Exposition, Inc.
Sulphur Springs—Hopkins County Fall Festival
and Fair

Tyler—East Texas State Fair
Waco—Heart O' Texas Fair

UTAH

Farmigton—Davis County Fair
Manti—Sanpete County Fair
Ogden—Weber County Fair
Salt Lake City—Utah State Fair

VERMONT

Barton—Orleans County Fair
Essex Junction—Champlain Valley Exposition
Guilford—Guilford Fair, Inc.
Lyndonville—Caledonia County Fair
Rutland—Vermont State Fair
St. Albans—Vermont Maple Festival
Council, Inc.
Tunbridge—Union Agricultural Society, Inc.
Winhall—The Bondville Fair

VIRGINIA

Abingdon—Washington County Fair and
Barley Festival
Arlington—Arlington County Fair, Inc.
Berryville—Clarke County Fair
Chase City—South Central Fair
Chesterfield—Chesterfield County Fair
Courtland—Franklin-Southamption
County Fair
Dublin—New River Valley Fair
Farmville—Five County Fair Association
Fredericksburg—Fredericksburg
Agricultural Fair
Front Royal—Warren County Fair
Harrisonburg—Rockingham County Fair
Isle of Wight—Isle of Wight County Fair
Lebanon—Russell County Fair
Lexington—Rockbridge Regional Fair
Madison—Madison County Fair
Manassas—Prince William County Fair
New Castle—Craig County Fair
North Garden—Albemarle County Fair, Inc.
Orange—Orange County Fair
Richmond—State Fair of Virginia
Ringgold—Danville-Pittsylvania County Fair
Salem—Salem Fair and Exposition
Saltville—Rich Valley Fair
South Boston—Halifax County Fair
Association, Inc.
Stuart—Patrick County Agricultural Fair
Tazewell—Tazewell County Fair
Warsaw—Richmond County Fair
Winchester—Frederick County Fair
Wise—Virginia-Kentucky District Fair
Woodstock—Shenandoah County Fair

WASHINGTON

Bremerton—Kitsap County Fair
Cashmere—Chelan County Fair
Centralia-Chehalis—Southwest
Washington Fair
Colfax—Palouse Empire Fair
Davenport—Lincoln County Fair
Ellensburg—Kittitas County Fair
Elma—Grays Harbor County Fair
Enumclaw—King County Fair
Graham—Pierce County Fair
Grandview—Yakima Valley Junior Fair
Kennewick—Benton-Franklin County Fair
& Rodeo
Lacey—Thurston County Fair
Langley—Island County Fair
Lynden—Northwest Washington Fair
Monroe—Evergreen State Fair
Moses Lake—Grant County Fair
Mt. Vernon—Skagit County Fair
Okanogan—Okanogan County Fair
Port Angeles—Clallam County Fair
Port Townsend—Jefferson County Fair
Puyallup—Puyallup Spring Fair
Puyallup—Western Washington Fair
Shelton—Mason County Fair
Spokane—Junior Livestock Show of Spokane
Spokane—Spokane Interstate Fair
Stanwood—Stanwood-Camano
Community Fair
Vancouver—Clark County Fair
Walla Walla—Walla Walla Fair & Frontier Days
Waterville—North Central Washington
District Fair
Yakima—Central Washington State Fair

WEST VIRGINIA

Danville—Boone County Fair Association
Eleanor—Putnam County Fair
Fort Ashby—Mineral County Fair
Lewisburg—State Fair of West Virginia
Middlebourne—Tyler County Fair
Milton—Cabell County Fair
Mineral Wells—West Virginia Interstate Fair
& Exposition
Moundsville—Marshall County Fair
New Martinsville—Town and County Days
(Wetzel County Fair and Festival)
Philippi, Barbour County Fair
Point Pleasant—Mason County Fair
Summersville—Nicholas County Fair
Association, Inc.
Sutton—Braxton County Fair
West Union—Doddridge County Fair
Association

WISCONSIN

Antigo—Langlade County Youth Fair
Beaver Dam— Dodge County Fair
Black River Falls—Jackson County Agricultural
   Society
Bloomington—Blake's Prairie Junior Fair
Chilton—Calumet County Fair
Chippewa Falls—Northern Wisconsin State Fair
Darlington—Lafayette County Fair
De Pere—Brown County Fair
Eau Claire—Eau Claire County Junior Fair
Elkhorn—Walworth County Fair
Ellsworth—Pierce County Fair
Elroy—Elroy Fair, Inc.
Fond du Lac—Fond du Lac County Fair
Friendship—Adams County Fair
Gays Mills—Crawford County Fair
Glenwood City—St. Croix County Fair
Grantsburg—Burnett County Agricultural Fair
Green Lake—Green Lake County Jr. Free Fair
Hayward—Sawyer County Agricultural Fair
   Association
Iron River—Bayfield County Fair
Janesville—Rock County 4-H Fair
La Crosse—La Crosse County Agricultural
   Society
Lancaster—Grant County Fair
Lodi—Lodi Agriculture Fair
Luxemburg—Kewaunee County Fair
Madison—Dane County Fair Association
Manitowoc—Manitowoc County Expo
Marshfield—Central Wisconsin State Fair
   Association, Inc.
Mauston—Juneau County Agricultural Society
Medford—Taylor County Fair
Merrill—Lincoln County 4-H Fair
Milwaukee (West Allis)—Wisconsin State Fair
Mineral Point—Iowa County Fair
Monroe—Green County Agricultural &
   Mechanics Institutional Fair
Oconto—Oconto County Youth Fair
Oshkosh—Winnebago County Fair
Phillips—Price County Fair
Plymouth—Sheboygan County Fair
Portage—Columbia County Fair Association
Rhinelander—Oneida County Fair
Rice Lake—Barron County Fair
St. Croix Falls—Polk County Fair
Seymour—Outagamie County Fair
Shawano—Shawano County Fair
Slinger—Washington County Junior Fair
Spooner—Washburn County Fair
Stoughton—Stoughton Junior Fair
Superior—Head of the Lakes Fair, Inc.
Tomah—Monroe County Fair

Viroqua—Vernon County Fair
Waukesha—Waukesha County Fair
   Association, Inc.
Wausau—Wisconsin Valley Fair
Wausaukee—Marinette County
   Fair Association
Wautoma—Waushara County Fair
Webster—-Central Burnett County Fair
Westfield—Marquette County Fair
Weyauwega—Waupaca County Fair
Wilmot—Kenosha County Fair

WYOMING

Afton—Lincoln County Fair
Basin—Big Horn County Fair
Big Piney—Sublette County Fair
Casper—Central Wyoming Fair and Rodeo
Douglas—Wyoming State Fair
Evanston—Uinta County Fair
Gillette—Campbell County Fair
Jackson—Teton County Fair
Laramie—Albany County Fair
Powell—Park County Fair
Riverton—Fremont County Fair
Rock Springs—Sweetwater County Events
   Complex
Sheridan—Sheridan County Fair
Torrington—Goshen County Fair and Rodeo
Wheatland—Platte County Fair and Rodeo
Worland—Washakie County Fair

CANADA

ALBERTA

Calgary—Calgary Exhibition & Stampede
Camrose—Camrose Regional Exhibition
Edmonton—Edmonton's Klondike Days
   Exposition
Lethbridge—Lethbridge & District
   Exhibition
Medicine Hat—Medicine Hat Exhibition
   & Stampede
Olds—Mountain View County Fair
Red Deer—Westerner Exposition Association

BRITISH COLUMBIA

Abbotsford—Central Fraser Valley Fair
   Association
Armstrong—Interior Provincial Exhibition
Cloverdale—Cloverdale Rodeo & Exhibition
Port Alberni—Alberni District Fall Fair
Vancouver—Pacific National Exhibition
Victoria—Victoria Exhibition

MANITOBA

Brandon—Provincial Exhibition of Manitoba
Winnipeg—Red River Exhibition

NEW BRUNSWICK

Chatham—The Miramichi Agricultural
   Exhibition Association, Ltd.
Fredericton—Fredericton Exhibition, Ltd.

NOVA SCOTIA

Halifax—Atlantic Winter Fair

ONTARIO

Barrie—Barrie Fair
Collingwood—Collingwood Great Northern
   Exhibition
Delta—Delta Agricultural Fair
Gloucester—Gloucester Agricultural Society
London—Western Fair
Markham—Markham Fair
Ottawa—Central Canada Exhibition
Simcoe— Norfolk County Fair
Stratford—Stratford Fall Fair
Thunder Bay—Canadian Lakehead Exhibition
Toronto—Canadian National Exhibition
Toronto—Royal Agricultural Winter Fair

QUEBEC

Quebec City—Expo Quebec

SASKATCHEWAN

Lloydminster—Lloydminster Agricultural
   Exhibition Association
Melville—Melville & District Agri-Park
   Association
Moose Jaw—Moose Jaw Exhibition
North Battleford—Battlefords' Exhibition
   Association
Regina—Canadian Western Agribition
Regina—Regina Buffalo Days Exhibition
Saskatoon—Saskatoon Praireland Exhibition
Swift Current—Swift Current Agricultural and
   Exhibition Association

Credits

I would like to thank Nina D.Hoffman and William R. Gray for publishing this, my first book. I am grateful to Barbara Payne, and to Coleen O'Shea, who defied both geography and time to coordinate this project. I also thank Dale-Marie Herring for her research efforts; Mark A. Mastromarino, whose expertise on county fairs was of great help; Lynne Warren for writing the legends; and Lisa Billard for her hard work and design. A big thanks goes to Randy Olson for his kindness and companionship—not to mention for driving me around the remoter parts of Iowa and Alabama. I am grateful to all the fair managers and volunteers who make county fairs possible each year. In addition, I thank the many people at the NATIONAL GEOGRAPHIC MAGAZINE who have given me so much support over the years: Bill Graves, Bill Allen, Betsy Moize, Erla Zwingle, Jennifer Reek, John Mitchell, Carol Lutyk, Ollie Payne, Joel Swerdlow, and Charlene Valeri. I offer my special thanks to Bob Poole. Finally, I would like to thank Amin Gulgee, whose belief in me has carried me through this and everything else.

John McCarry

I would like to thank the National Geographic Society, which has been helpful and supportive of my work, especially Nina D. Hoffman and William R. Gray for publishing this book. In addition, Barbara Payne, Bill Allen, Tom Kennedy, Kent Kobersteen, Kathy Moran, Bruce McElfresh, and Charles Kogod, for their continued support. I also thank Lynne Warren for her work on the legends; Lisa Billard for her design; and Coleen O'Shea, project director. Thanks to John McCarry, my companion on the Ejection Seat ride at Iowa's Clay County Fair. I extend particular thanks to the many county agents, fair workers, and farm families who were an essential part of this book. And special thanks to my wife, Melissa Farlow, for her patience and help in editing my work.

Randy Olson

Published by The National Geographic Society
Reg Murphy *President and Chief Executive Officer*
Gilbert M. Grosvenor *Chairman of the Board*
Nina D. Hoffman *Senior Vice President*
William R. Gray *Vice President and Director, Book Division*
Charles Kogod *Assistant Director*
Barbara A. Payne *Editorial Director and Managing Editor*
Dale-Marie Herring *Assistant to the Editorial Director and Project Reseacher*
Lynne Warren *Picture Legends Writer*
Mark A. Mastromarino *Text Consultant*
Richard S. Wain *Production Project Manager*
Manufacturing and Quality Management
George V. White *Director*
Vincent P. Ryan *Manager*

Prepared for the National Geographic Society by
Project Director *Coleen O'Shea Literary Enterprises*
Design *Lisa Billard Design, NY*